METALS AT HIGH TEMPERATURES

METALS AT HIGH TEMPERATURES
Standard Handbook of Properties

V. E. Zinov'yev

Translated and Edited by

V. P. Itkin
University of Toronto

In coordination with the

National Standard Reference Data Service of the USSR

◯HEMISPHERE PUBLISHING CORPORATION
A member of the Taylor & Francis Group
New York   Washington   Philadelphia   London

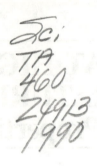
METALS AT HIGH TEMPERATURES: Standard Handbook of Properties

Originally published as Kineticheskie svoystva metallov pri vysokikh temperaturakh by Metallurgiya, Moscow, 1984.

1 2 3 4 5 6 7 8 9 0   B C B C   9 8 7 6 5 4 3 2 1 0

This book was set in Times Roman by Edwards Brothers, Inc. The editor was Eugene Drachner. Printing and binding by BookCrafters, Inc.

A CIP catalog record for this book is available from the British Library.

Library of Congress Cataloging-in-Publication Data

Zinov'yev, V. E. (Vladislav Eugen'evich)
    [Kineticheskie svoĭstva metallov pri vysokikh temperaturakh. English]
    Metals at high temperatures : standard handbook of properties /
V. E. Zinov'yev ; translated and edited by V. P. Itkin.
        p. cm.
    Translation of: Kineticheskie svoĭstva metallov pri vysokikh temperaturakh.
    Includes bibliographical references.
    1. Metals—Thermal properties. 2. Metals at high temperatures.
3. Matter, Kinetic theory of. I. Itkin, V. P. II. Title.
TA460.Z4913   1990
620.1′696—dc20   90-4289
ISBN 0-89116-853-2   CIP

CONTENTS

To understand and predict the behavior of materials at high temperatures, it is necessary to know their physical properties. However, until recently, much of the high temperature data on the kinetic and thermophysical properties of many construction materials, and even pure metals, was unreliable. In many cases, this was due to the difficulties of conducting high temperature experiments, and in some cases even qualitative results could not be obtained. The development of techniques of high temperature thermophysical experiments over the last 10–15 years helped to overcome these difficulties.

Summarizing the points mentioned above, it can be said that the data available in the reference books require corrections and additions. The present author attempted to provide complete information on electrical and thermal conductivities, thermal diffusivity, thermoelectric power, and the Hall coefficient of metals for a broad temperature interval (usually starting from 100 K, in some cases from 4.2 K, and up to the melting point, and in some cases for the liquid state as well). In addition to these characteristics, the parameters derived from the velocity of sound, that is, the elastic constants, and the moduli of elasticity are given over a broad temperature interval. Although the information reviewed here applies mainly to the high temperature region (in physics, temperatures above the Debye temperature are considered high, i.e., 100–400 K), sometimes, for reasons of completeness, the data are given for the low and moderate temperatures.

Several tables (see Tables 6, 14, 16, 17, 20–23, 25–28, 30, 32, 36, 41, and 46) list the values which were statistically analyzed by the present author. An attempt was made in this book to discuss the reviewed and systematized experimental data from the position of modern solid-state physics. Special attention was paid to the transition metals, for which a large quantity of new information was recently obtained. The principal mechanisms of electron and phonon scattering were isolated from the analysis of data on the kinetic properties of metals. This may be useful for a theoretical analysis of the transport phenomena in metals.

The author wishes to cordially thank P. V. Gel'd for his support, and a number of scientists from the Department of Physics at the Sverdlovsk Mining Institute, the Ural Polytechnical Institute, and the Izhevsk Mechanical Institute whose results were used in this book: I. G. Korshunov, A. D. Ivliyeva, S. A. Il'inykh, L. I. Chupina, V. I. Sperelup, S. G. Taluts, and L. D. Zagrebin. The author wishes to express his gratitude to I. M. Sperelup and N. G. Gorbunova, who participated in the preparation of the text and illustrations for this book, and also to G. P. Zinov'yeva, who prepared the materials on the elastic constants of metals.

The author would be glad to receive any comments or suggestions regarding this book.

V. E. Zinov'yev

INTRODUCTION

It is customary to use the term "kinetic properties" for such characteristics of metals that are determined by the parameters of the transport of energy, charge, and mass. In condensed substances like metals, at the microscopic level, the energy and charge are transported by electrons and phonons, and the most important kinetic properties are electrical conductivity, thermal conductivity, and diffusivity. These simplest characteristics of metals are followed by many thermoelectric, galvanomagnetic, and thermogalvanomagnetic characteristics [1, 2]. Among them, the thermoelectric power and the isothermal Hall coefficient are the most well-known properties. The above-mentioned characteristics of metals are primarily determined by the electronic subsystem. The parameters, determined from the velocity of sound in metals, are related to the transport properties of the lattice subsystem. These parameters are the elastic moduli and the elastic constants of a crystal, and they are very often called the mechanical characteristics of the crystal lattice, despite the fact that they are obtained from the measurements of sound wave propagation.

The absence of reliable experimental data considerably hinders the development of theoretical postulates concerning the high temperature kinetic properties of metals.

The intensive development of the experimental technique for measuring the thermal kinetic properties, electrical resistivity, thermoelectric power, velocity of the propagation of sound, and success in obtaining pure single crystal metals increased considerably the quantity of reliable experimental data. The successful contributions from the Soviet school of thermophysics should be mentioned.

As the author has some experience in investigating the kinetic properties of metals at high temperatures, the selection of the most reliable data was made according to a certain set of criteria. The experimental procedure had to be based on reliable and tested methods and the samples of metals being used must have been the purest possible, preferably single crystals. Where possible, the recommended data were obtained from a statistical analysis of numerous experimental data. For other cases, the given data were obtained from averaging the data of several investigations (judged by the author to be the most reliable), or included the results of individual investigations. For all of the cases, the given uncertainty of the results comes from the corresponding literature. The uncertainty in the values of the absolute thermoelectric power and the Hall coefficient is rarely available; the experimental uncertainty of the Hall coefficient is guessed to be about 10–15%. The uncertainty in the given values of the absolute thermoelectric power is harder to determine, because very often the values cross the horizontal axis (change in sign). Experience in this area shows that the basic shape of the temperature dependence of thermoelectric power can be reproduced quite well, but the position of the horizontal axis can fluctuate within several microvolts. This can considerably change the positions of the inversion points, and the problem probably requires a special investigation. The lack of available experimental information concerning the physical properties of metals is reflected in the tables, where hyphens indicate the absence of reliable data.

In conclusion, the acquisition of reliable transport coefficients of metals, keeping the uncertainty to only several percent, still remains an unresolved problem, especially for metals with a noncubic lattice.

INFORMATION ON BASIC KINETIC PROPERTIES OF METALS

Several monographs and textbooks have been dedicated to the topic of transport phenomena in metals. For example, [1] and [2] can be recommended as good references. The majority of published results usually apply to the low temperature region. At the present time the degree of theoretical development is not so high as to allow the quantitative calculations of kinetic coefficients for even the ordinary alkali metals, not to mention the more complex transition metals. Very often the theoretical calculations are limited to explanation of the temperature dependence of the electrical resistivity. Attempts to describe all kinetic properties within a single model are rare.

In general, the analysis of the kinetic properties of metals at high temperatures requires a complex model with many-particle interactions, and heavy computing power. Significant difficulties in developing such a model make it necessary to find simple, even rough, models capable of adequately describing the properties of groups of metals for limited temperature ranges.

§1 DETERMINING TRANSPORT COEFFICIENTS

The transport coefficients N_{ik} are determined from the transport equations for the electric current density $\vec{j}$ and heat flux density $\vec{\omega}$ [1]:

$$j = N_{11} \operatorname{grad} \eta/e + N_{12} (-\operatorname{grad} T/T), \quad \omega = N_{21} \operatorname{grad} \eta/e + N_{22} (-\operatorname{grad} T/T) \qquad (1)$$

where η is the electrochemical potential; T is the temperature; and, e is the charge of the electron. If a magnetic field is absent, the values of N_{ik} are determined relatively simply from the experimentally found parameters. Then the equations (1) can be presented in the following form

$$\text{grad } \eta/e - j/\sigma + S \text{ grad } T$$

$$\omega = P\vec{j} - \lambda \text{ grad } T \tag{2}$$

where $\sigma = N_{11}; \quad S = N_{12}/(TN_{11})$

$$\lambda = (N_{11} N_{12} - N_{12} N_{21})/(N_{11} T); \quad P = N_{21}/N_{11} \tag{3}$$

If grad $T = 0$, then $\vec{j}/\sigma \text{grad } \eta/e = \text{grad } (\zeta/e - \varphi)$, where ζ is the chemical potential and φ is the electric field potential. For a homogeneous solid body, the electric field intensity can be expressed as $\vec{E} = -\text{grad } \varphi$. In this case, the proportionality coefficient between $\vec{j}$ and $\vec{E}$ is the electrical conductivity, σ; its inverse is the electrical resistivity: $1/\sigma = \rho; \vec{E} = \rho\vec{j}$.

It follows from eq. (2) that $\vec{\omega} = P\vec{j}$, where P is the coefficient of proportionality between the electric current and the thermal energy transfered by it. For the experimental determination of the coefficient P, the additional heat emission or absorption is considered. It is caused by the contribution of $\vec{j}$ grad P against the Joule heat $j^2 \rho$. In homogeneous metals P is a constant, and grad $P = 0$. However, at a point where two different metals, A and B, make contact, the coefficient P is a constant for the two given metals. The heat being released, $\vec{j}$ grad P, is called the Peltier effect, and P is known as the Peltier coefficient.

If $j = 0$, and grad $T \neq 0$, than grad $(\eta/e) = S$ grad T and $\vec{\omega} = -\lambda$ grad T. Here λ is the thermal conductivity, and S is the absolute differential thermoelectric power.

If the soldered terminals of the two conductors connected in series, A and B, are held at different temperatures, T_1 and T_2, then thermoelectric power appearing between the end point terminals can be expressed as

$$\delta\varphi = \int_{T_2}^{T_1} (S_A - S_B) dT$$

The following consideration is restricted by properties of homogeneous conductors only. It should be noted that for homogeneous anisotropic conductors, the coefficients σ, λ and S are tensors.

If in a homogeneous metal with the electric current flowing in the direction of x, the external magnetic field, B_z, is applied perpendicularly to the current direction, then a Hall field, E_y, is produced in a direction perpendicular to both the current and external field. The corresponding coefficient is called the Hall coefficient: $R = E_y/j_x B_z$.

The Hall coefficient is usually measured at a constant temperature. Its value depends on the magnetic field intensity. There are different Hall coefficients for weak and strong magnetic fields.

§2 MECHANISMS OF ELECTRON AND PHONON SCATTERING IN METALS AT HIGH TEMPERATURES

The concepts used for the analysis of transport phenomena in metals are often based on singling out the most specific charge and energy carriers, and mechanisms of their scattering on different imperfections of a crystal lattice. As a first approximation, it is assumed that different mechanisms are independent of each other, and that their contributions to the total resistance or flux are additive. A more detailed approach requires taking into account their interactions. The separation and analysis of contributions of different mechanisms to the kinetic coefficients are the principle topics in the kinetic theory of metals and alloys; monographs [1–3], for example, discuss these topics. The principal concepts, used at the present time in the analysis of the kinetic properties of metals at moderate and high temperatures, are considered below.

ELECTRICAL RESISTANCE

It is known that resistance to the electric current is due to distortions of the translational symmetry of a crystal lattice. According to the nature of these distortions, the electrical resistivity of metals can be represented as a sum

$$\rho = \rho_i + \rho_{e-ph} + \rho_{e-e}$$

where ρ_i, ρ_{e-ph} and ρ_{e-e} are contributions caused by the scattering of electrons by static imperfections, phonons and electrons, respectively.

Scattering by static imperfections. Static imperfections can be one-dimensional (vacancies, self-interstitual atoms, interstitial and substitution atoms of impurities), two-dimensional (dislocations) and three-dimensional (grain and crystal boundaries, vacancy clusters).

If the concentration of point defects is low (not more than several percent), the additional resistivity, which is caused by imperfections, is approximately proportional to their concentration [2, 3]

$$\rho = \rho_M + \Delta\rho \approx \rho_M + Ac \tag{4}$$

where ρ_m is the resistivity of a matrix; $\Delta\rho$ the additional resistivity; c the concentration of point defects; and A is a constant which depends on a scattering cross section of point defects.

In a pure metal approaching the melting point, the influence of equilibrium thermal vacancies can be observed. Their concentration depends on temperature, and it can be described by the relation $c = B \exp(-E_v/kT)$, where E_v is the energy of vacancy formation, and k is the Boltzmann constant.

The effect of vacancies was analysed in studies [4–6], where it was pointed out, that near the melting point the value of c may range from several hundredth of a percent to several percent, and $\Delta\rho$ to several $\mu\Omega\cdot$cm.

The scattering cross section of point defects depends on the value of z, the difference between the valences of matrix and impurity atoms. Therefore the additional resistivity can be expressed [2, 3] as

$$\Delta\rho = a + bz \tag{5}$$

where a and b are the constants for a given metal and row in the periodic table, where the impurity atom is located. Equation (5) is called Linde's rule [2, 3].

In the limited temperature interval, the value of A in eq. (4) does not depend on temperature [2, 3]. Therefore, the value of $\partial\rho/\partial T$ does not depend on the impurity concentration

$$d\rho/dT = (d\rho_{e\text{-}ph}/dT) + (d\rho_{e\text{-}e}/dT)$$

For substitutional solid solutions of A and B metals, the additional resistivity is taken into account, relatively often, through a parabolic law [2, 3] $\Delta\rho \approx C(1 - C)$, where C and $(1 - C)$ are mole fractions of the elements A and B, respectively.

It should be noted, that the value of the residual electrical resistivity of metals $\rho_i \approx \rho_{4.2K}$ may serve as one of the purity characteristics. The ratio of the electrical resistivity at room and helium temperatures, $r = \rho_{293K}/\rho_{4.2K}$ is another and more frequently used characteristic of metal purity. If the temperature of the transition to the superconductive state, T_c, is above 4.2 K, then $\rho_i \approx \rho_{T_c}$, where ρ_{T_c} is the electrical resistivity at temperatures near T_c in the normal state. For sufficiently pure metals, this ratio is of the order of 10^2, and for perfect single crystals it may reach 10^5. The magnitude of the impurity contribution to the electrical resistance at moderate and high temperatures is often of the same order as the uncertainty in the determination of the value of ρ. Therefore, the influence of impurities at temperatures above 200–300 K can be neglected. The last part of the above statement does not apply to the region of phase transformations where the nature of $\rho(T)$ is influenced more by impurities, especially near the transformation points.

The contribution of dislocations to the electrical resistivity is temperature independent and determined by $\rho_d = AN_d$, where N_d, is the concentration of dislocations and A is a coefficient having values between $5\cdot10^{-21}$ and $2\cdot10^{-19}$ $\Omega\cdot$cm.

It is convenient to study the influence of a plastic deformation on the resistance at low temperatures. At moderate and high temperatures, the contributions of static imperfections in sufficiently pure crystals can often be neglected. These contributions amount to no more than one percent of the total resistance. However in sensitive experiments, and especially near the phase transformation points, the influence of static imperfections becomes noticeable.

Scattering by phonons. This scattering mechanism is considered in an adiabatic approximation. This method treats the electronic states of the system as being independent of the vibration states of the lattice. Therefore, every subsystem gives a separate independent contribution to the total energy [2]. In this approximation, the wave functions of electrons depend on the instant coordinates of ions. The adiabatic approximation is applicable when $h/kT < \tau$, where τ is the relaxation time of an electron. For metals at high temperatures, this condition is not met very well. However, [2] showed that for such a kinetic phenomenon as electrical conduction, the above condition can be replaced by a less strict expression $(v_s/v_F)\ (h/kT) < \tau$, where v_s and v_F are velocities of sound and Fermi electrons, respectively. The uncertainty principle is responsible for an important condition, which restricts the regular approximations used in the theory of the transport phenomena

$$1 < k_F \Lambda_e \tag{6}$$

where k_F is the modulus of a wave vector of the Fermi electron, and Λ_e is the length of the mean free path.

This relation is called Landau-Peierls criterion [2]. According to [8], this relation can significantly influence the kinetic properties of the transition metals at high temperatures.

At moderate and high temperatures the interaction of electrons with the lattice is considered, taking into account that the longitudinal phonons are the main participants of regular scattering processes. The contribution of transverse phonons may be noticeable, either in metals with Fermi surfaces (which are significantly different from spherical) or in the Umklapp processes. The matrix element of scattering depends highly on the angle of scattering, so that at high and moderate temperatures the wide-angle electron scattering ("90 degree scattering") plays the most important role.

The electrical conductivity is found from the general relation for the current density

$$\vec{j} = \int e v_\kappa \vec{f}_\kappa \, d\vec{k}$$

where $\vec{f}_{\vec{k}}$ is the distribution function of the electrons in the space of wave vector $\vec{k}$, and $\vec{v}_{\vec{k}}$ is the velocity of the electrons. For the determination of the kinetic coefficients, the relaxation time τ is often used. This is the time interval between two consecutive collisions of electrons with defects in the crystal lattice or the time between the electron-phonon transitions [1, 2]. In general, the relaxation time is anisotropic and varies for different kinetic phenomena [1, 2]. The electrical conductivity is described by the relaxation time as

$$\sigma = (e^2/4\pi^3 h) \int_{S_F} \tau\left(\vec{\kappa}\right) \vec{v}_\kappa \, d\, \vec{S}_F \tag{7}$$

where S_F is the area of the Fermi surface.

At high temperatures (above the Debye temperature) the concept of isotrope relaxation time is often used. Then the expression for the electrical conductivity appears as

$$\sigma = (e^2 \tau / 4\pi^3 \hbar) \int_{S_F} \vec{v} d\vec{S_F} \tag{8}$$

For anisotropic noncubic crystals, the anisotropy in electrical conductivity along the principal crystallographic directions will be determined only by the anisotropy of the Fermi characteristics. Particularly, for a hexagonal close-packed lattice

$$\rho_\perp / \rho_\parallel = \int_{S_F^\parallel} v_F dS_F / \int_{S_F^\perp} v_F dS_F \tag{9}$$

where $S_F^\perp$ and $S_F^\parallel$ are the respective projections of the Fermi surface on the planes perpendicular and parallel to the hexagonal axis.

The simplest approximation, used for calculating the temperature dependence of the electrical resistivity of metals, is the Debye model for a phonon spectrum and the model of a spherical Fermi surface. The expression for the electrical resistivity of cubic metals, which was obtained using these approximations, is known as the Bloch-Gruneisen formula

$$\rho = 4\rho_{\theta_D}(T/\theta_D)^5 I_5(\theta_D/T) \tag{10}$$

where ρ_{θ_D} is a certain constant for a given metal, and I_5 is the kinetic integral of the type

$$I_5(\theta_D/T) = \int_0^{\theta/T} \{x^5 dx / [(e^x - 1)(1 - e^{-x})]\}$$

The constant ρ_{θ_D} can be expressed as

$$\rho_{\theta_D} = \pi^3 \hbar^3 R_D / (4e^2 Mzk\theta_D) \tag{11}$$

where M is the mass of an ion; z is the number of free electrons per atom; and, $R_D = (3\pi^2\hbar)^{1/3}$ is the radius of the Debye sphere [1, 2].

In the low temperature region, the value of $I_5 \theta_D T^{-1} \rightarrow 124.4$, so that the low-temperature phonon-electron electrical resistivity is $\rho = 497.6 \rho_{\theta_D} (T/\theta)^5$.

For the high temperature region

$$I_5(\theta/T) \approx \frac{1}{4}(\theta/T)^4; \quad \rho = \rho_{\theta_D}(T/\theta_D)$$

i.e. at $\theta_D = $ const, the resistivity temperature dependence is linear.

The coefficient ρ_{θ_D} is slightly temperature dependent because of the thermal expansion of crystals. This thermal expansion is taken into account as $\theta_D = \theta_D^0 (1 - \beta\gamma T)$, where θ_D^0 is the Debye temperature at $T = 0$; β is the temperature coefficient of volume expansion and γ is the Gruneisen constant. The decrease in the Debye temperature results in an increase in ρ.

Inter-band electron-phonon scattering in transition metals. In transition metals a special mechanism of inter-band scattering was proposed by Mott and investigated in detail by Wilson [9, 10]. The mechanism describes the scattering of s-electrons into the d-band of the transition metals. Since the d-band is rather narrow and has a high density of electronic states, the light s-electrons may occupy the vacant states of the d-band after colliding with the phonons. As a result, the effective mass of electrons increases sharply, and the total electrical resistance is also increased.

The method for calculating the kinetic properties of the transition metals at high temperatures was derived in [11] where the so-called method of moments was used. It is a convenient semiempirical method to determine the parameters of the band structure and the specific mechanisms of scattering on the basis of certain experimental data [12]. In this method the product τv^2 is averaged over equal energy surfaces, and the value of energy dependent electrical conductivity $\sigma(\varepsilon)$ is introduced:

$$\sigma(\varepsilon) = N^3(\varepsilon)\, \tau(\varepsilon)\, m^{*-1}$$

where $N(\varepsilon)$ is the density of electronic states as a function of energy and m^* is the effective mass.

It is assumed that the value of $\sigma(\varepsilon)$ determines the principal transport coefficients, and that it is identical for all of them. This assumption is equivalent to approximating the relaxation time. Using eq. (3), the electrical conductivity can be derived [2, 12]

$$\sigma_{s\,d} = c^2 M_0$$

$$M_n = -\int_{-\infty}^{+\infty} \frac{\partial f_0}{\partial \eta} \eta^n \sigma(\eta)\, d\eta \tag{12}$$

where n is an integer (for the electrical conductivity); $n = 0$, $\eta = (\varepsilon - E_F)/(RT)$.

Study [11] analyzed the properties of paramagnetic transition metals and concluded that these metals may be divided into 'minus' and 'plus' groups, corresponding to the temperature dependence of the magnetic susceptibility and electrical conductivity. These groups alternate in the periodic system. Metals of the scandium, vanadium, manganese, palladium and platinum subgroups belong to the 'minus' groups; metals of the titanium, chromium, iridium and rhodium subgroups belong to the 'plus' groups. In a 'plus' group magnetic susceptibility increases with temperature, while in a minus 'group' it decreases. As a rule, for the 'minus' groups $\partial^2 \rho / \partial T^2 > 0$, and for the 'plus' groups $\partial^2 \rho / \partial T^2 < 0$.

These features of the kinetic properties are explained by the densities of the electronic states. These densities are high for the metals of the 'minus' groups, for which the Fermi levels are located near the maximum of $N(\varepsilon)$.

Near the Debye temperature the expression (12) can be expanded

$$\rho_{s\,d} = AN(\varepsilon)\, \upsilon_D^{-2}\, T\, \{1 - [\pi^2/6\,(kT)^2\,(3N_1^2 - N_2)]\} \tag{13}$$

$$N_1 = \frac{1}{N(\varepsilon)}\frac{\partial N(\varepsilon)}{\partial \varepsilon}\bigg|_{\varepsilon=E_F}; \quad N_2 = \frac{1}{N(\varepsilon)}\frac{\partial^2 N(\varepsilon)}{\partial \varepsilon^2}\bigg|_{\varepsilon=E_F} \tag{14}$$

Electron-electron coulomb scattering. Study [2] showed that the regular electron-electron processes make no contribution to the electrical resistivity. However, the electron-electron contribution may be observed in ordinary metals, in the Umklapp processes, as a result of the Coulomb scattering. At room temperature the relative magnitude of such a contribution is on the order of 0.1%, making it unnoticeable among the other contributions. At low temperatures $\rho_{e-e} \sim T^2$. The electron-electron contributions can be observed against the background of a phonon contribution $\rho_{e-ph} \sim T^5$. However this effect has not been detected experimentally [2].

In transition metals, the processes of interband electron-electron scattering are possible. These processes are accompanied by changes in the effective masses of the carriers [2].

According to the data of [2, 13], in transition metals this contribution is higher than in ordinary metals by approximately $(v_s/v_d)^2$ times, where v_s and v_d are the velocities of the light s and heavy d carriers, respectively. A term in the expression for the electrical resistivity of the transition metals, which is quadratic in temperature, $\rho_{e-e} = BT^2$, is easily visible at hydrogen and helium temperatures and can exceed substantially the phonon contribution [14]. Usually this term is difficult to observe against the background of other contributions at moderate and high temperatures.

Electron scattering by magnetic inhomogeneities. In metals that are magnetically ordered at low temperatures, the scattering resulting from the distortion of the magnetic order is very substantial. Two simple approaches are generally used to describe this phenomenon, with Mott's band mechanism being one of them. It is based on the notion that a narrow band of d-holes splits in the internal magnetic field. The resulting sub-bands of spins-up and -down shift relative to each other during the magnetic ordering [2, 15]. Such a polarization produces a change in the densities of states (of d-holes), into which the s-electrons can be scattered. The simplest considerations show that the resistivity ρ_m, arising from this mechanism, will be defined by the magnetization.

$$\rho_m = \rho_{s-d} \frac{1}{2} \left[\left(1 - \frac{M}{M_0} \right)^{1/3} + \left(1 + \frac{M}{M_0} \right)^{1/3} \right] \tag{15}$$

where M is the average magnetization which reaches a saturation value of $M = M_0$, at $T = 0$. Above the magnetic disordering point $M = 0$, the additional contribution to the resistance is similar to Mott's term ρ_{s-d} in Eq. (15). Below T_c the magnetic ordering leads to a decrease in resistance, in comparison with the ordinary s-d-contribution. The temperature dependence of resistivity in metals that are in a magnetically ordered state has a characteristic shape; $\partial^2\rho/\partial T^2 > 0$ below T_c, and $\partial^2\rho/\partial T^2 < 0$ above T_c, with the inflection point near T_c.

Another way to describe the properties of magnetically ordered metals is based on a one band model. This model assumes that the s-electrons, which form a broad band, are scattered by the magnetic moments of the localized d-electrons. This mechanism was proposed initially in the studies of Kasuya and de Jeun and Fridel.

At high temperatures, below the Curie point, the expression for the additional resistivity can be described as

$$\rho_m = A\left[\frac{1}{2} <(s+S_t)(s-S_t+1)> \frac{R_M}{kT}\left(\exp\frac{R_M}{kT}-1\right)^{-1}\right.$$
$$\left.+\frac{1}{2}<(s-S_t)(s+S_t+1)>\frac{R_M}{kT}\left(1-\exp\frac{R_M}{kT}\right)\right] \tag{16}$$

where A is a constant defined by the universal constants and parameters of the crystal; s is the maximal projection of the spin of a crystal lattice point; S_t is the projection of the spin in the direction of the magnetic field; and, R_m is a constant of the molecular field. In the ferromagnetic region distant from T_c, $R_m \to 0$, Eq. (16) gives only a small contribution to total resistivity.

The magnitude of the magnetic contribution increases rapidly as the temperature approaches T_c, when the slope of the $\rho(T)$ dependence is positive, and when $T = T_c$

$$\rho_m = (A/2)\,s\,(s+1) = D \tag{17}$$

i.e., the value of ρ_m does not depend on the temperature. Thus, above the point of the magnetic disordering, the total resistivity of a pure metal has the following temperature dependence

$$\rho = \rho_{e\cdot ph} + \rho_m \approx AT + D \tag{18}$$

while $\partial\rho/\partial T = 0$ or slightly decreases. These expressions allow the separation of the magnetic and the phonon contributions. The question of which of these two discussed models better applies to the electron scattering requires, for each specific metal, an explanation which takes into account other kinetic and static properties of that metal. Quite often both types of scattering have to be taken into account.

Scattering by the magnetic inhomogeneities can also occur in non magnetic transition metals with a strong exchange interaction [16]. In these 'almost magnetic' metals (such as actinides, palladium and platinum), the energy of the magnetic interaction is not sufficient for creating a magnetic order. However, the strong internal magnetic interaction leads to a correlated interaction of spins (paramagnetic states are formed), the fluctuations of which cause additional spin scattering of the kinetic s-electrons. Study [16] showed that, at high temperatures, this mechanism degenerates into scattering by localized magnetic inhomogeneities in separate lattice sites. This results in an equation like (17), which is, apparently, related to a decrease in the paramagnon size. There are several variants of paramagnon scattering [16, 17]. For high temperatures the model suggested by Jullien [16] and Mills [17] is the most developed one. It assumes that the d-electrons form a narrow paramagnon band of $\sim kT$ width, and the s-electrons are scattered by the paramagnon states, giving a high temperature contribution of the type:

$$\rho_{e\cdot pm} = \rho_\infty\left[1 + 2/3\,(\chi_0 I - 4/3\xi^2)(T_{F_d}/T)\right] \tag{19}$$

where $\rho\infty$ is the resistivity value at 'saturation', i.e., at $T \to \infty$; χ is the magnetic susceptibility at $T = 0$; I is the so-called Stoner exchange factor; $\xi = k_{F_d}/k_{F_s}$ is the ratio of the Fermi wave numbers for the narrow d-band and the broad s-band; and, T_{F_d} is the Fermi temperature for the d-band. Strong nonlinearities in $\rho(T)$ are allowed by this model; in addition, if $\chi_0 I > \frac{4}{3}\xi^2$, than $\partial^2\rho/\partial T^2 < 0$. This happens for palladium and platinum and also for metals of the scandium subgroup. If $\chi_0 I < \frac{4}{3}\xi^2$, then $\rho(T)$ has a maximum point and, at high temperatures, a negative temperature coefficient. This is observed for several actinides.

Separation of different mechanisms of scattering on the basis of only one property is often impossible. Determining the nature of different scattering mechanisms requires a complex analysis of kinetic characteristics at the broadest possible temperature interval.

Scattering near the points of the magnetic disordering. A specific type of scattering of electrons is observed near the magnetically disordered points. Electrical resistivity has an inflection point in the vicinity of the Curie point. Therefore, a partial derivative of resistivity, with respect to temperature, has a λ-shaped maximum. In very pure metals, the behavior of the electrical resistivity in the vicinity of T_c has an even more pronounced anomaly: a cusp, so that the maximum in $\partial\rho/\partial T$ at T_c is accompanied by a jump (see the data of studies [18–24] for nickel, iron and cobalt). There are numerous theoretical studies discussing the anomaly in electrical resistivity in the vicinity of the Curie point [18–22]. The principal conclusion of these studies is that the anomaly in $\partial\rho/\partial T$ is proportional to the anomalous contribution to the specific heat

$$C_m \sim \frac{\partial\rho}{\partial T} = \begin{cases} A' + B' t^{-\alpha'}, & T < T_C \\ A + Bt^{-\alpha}, & T > T_C \end{cases} \tag{20}$$

where $t = (T - T_c)/T_c$ is the normalized temperature; α and α' are the so-called critical indices; and, A and B, and A' and B' are constants for the $T > T_c$ and $T < T_c$ parts of the graph.

Impurities affect the anomaly very strongly, making it less pronounced. The physical nature of the anomaly lies in the scattering of the conducting electrons by the fluctuations of the magnetic inhomogeneities. The short-range order fluctuations provide the main contribution [18–21].

Antiferromagnetic disordering can make the anomaly in resistivity even more pronounced, with a maximum point appearing even in $\rho(T)$ [21]. The nature of this anomaly is also based on critical scattering. However, in this case, the contribution of the long-range order fluctuations is significant [25].

THERMAL CONDUCTIVITY

Electronic and phonon mechanisms account for heat transfer in nonmagnetic metals: $\lambda = \lambda_e + \lambda_g$. In magnetically ordered metals, one more mechanism can be added. It involves heat transfer by the spin waves (or the magnetic subsystem)

λ_m. Finally, in semiconductors or semimetals, approaching semiconductors in electronic structure, another mechanism can exist. It is caused by the bipolar thermal diffusion λ_s. At $T > \theta_D$ the contributions from the last two mechanisms are small, and the main problem is how to separate the phonon and the electronic components of the thermal conductivity. Unfortunately, in metals at moderate and high temperatures, there are no direct experimental methods of separating these components. In pure metals at low temperatures, these components can be separated by a magnetic field. To perform such a separation at high temperatures, it is necessary to use a different theoretical approach. The Wiedemann-Franz-Lorenz (W-F-L) law is one of the major theoretical concepts that can be used for this purpose.

On the other hand, the phonon contribution to the thermal conductivity can be estimated independently, on the basis of the theory of lattice vibrations [2, 23]. Therefore, it is possible to estimate independently the electronic and phonon contributions and their temperature dependence. It should be stressed that the problem of correctly separating the various contributions to thermal conductivity for different metals remains the subject of research. This is the result of a weak theoretical foundation and of great complications in the physics and mathematics involved. To define qualitatively the relationship between the various contributions, it is necessary to use approximate methods of separation, based on the W-F-L law, and to calculate the value of λ_g for a possibly broader group of metals. Such estimations will be presented below in the example of the transition metals investigated by the present authors.

Considering separately every component of the thermal conductivity, the electronic thermal resistivity can be expressed

$$\lambda_e^{-1} = W_e = W_{e-i} + W_{e-ph} + W_{e-e}$$

where W_{e-i}, W_{e-ph}, and W_{e-e} are the thermal resistivities resulting from the scattering of electrons by impurities, phonons and electrons, respectively (including magnetic inhomogeneities). For the phonon thermal resistivity

$$\lambda_g^{-1} = W_g = W_{ph-i} + W_{ph-ph} + W_{ph-e}$$

where W_{ph-i}, W_{ph-ph}, and W_{ph-e} are the thermal resistivities resulting from phonon-impurity, phonon-phonon and phonon-electron scattering, respectively.

Even from these simple relations, it is clear that the mechanisms of scattering and conductivity are closely bound. When contributions to one of the mechanisms change, they may cause changes in contributions to other mechanisms. This problem will be discussed with more detail when the methods of separating the mechanisms of the transport phenomenon and thermal energy dispersion are considered for specific metals.

Lattice thermal conductivity. Lattice thermal resistivity is caused by scattering resulting from defects in the crystaline lattice and from impurities. In the first approximation it does not depend on temperature above $\theta_D/3$ [2, 12, 24].

The principal contribution to lattice thermal resistivity at moderate and high temperatures is caused by the three-phonon processes of scattering. The theory was developed, in detail, by Leibfried and Schlomann [2, 12, 23]. The main conclusion of this theory is that lattice thermal resistivity is proportional to the temperature $W_{ph-ph} = AT$. The proportionality coefficient may be estimated by several methods. A convenient approximation is the formula of Dugdale and MacDonald [23]

$$W_{ph\text{-}ph} = 9\alpha\gamma T/(\bar{v}_s\, c_V\, a) \tag{21}$$

where α is the coefficient of linear thermal expansion; $\bar{v}_s$ is the average velocity of sound $\bar{v}_s = \frac{1}{3}(v_l + 2v_t)$, where v_l is the velocity of longitudinal waves; v_t is the velocity of transverse waves; c_v is the specific heat capacity at constant volume; a is the average lattice parameter; and, γ is the Gruneisen parameter.

Phonon scattering by electrons also produces a contribution to the lattice thermal resistivity. An estimation of the value and temperature dependence of this contribution, performed in studies [2, 23], shows that at high temperatures

$$W_{e\text{-}ph} \approx (e/k)^2\, z_e^2\, (\rho_{e\text{-}ph}/T) \tag{22}$$

where z_e is the effective number of conduction electrons per atom; and, ρ_{e-ph} is the electron-phonon contribution to the electrical resistivity.

At high temperatures W_{e-ph} does not depend on temperature because $\rho_{e-ph} \sim T$. It should be noted that in transition metals W_{e-ph} may be of the same order of magnitude or even greater than W_{ph-ph} [2, 8, 23]

Table 1 lists the results from estimating the contributions to the thermal resistivity and conductivity of the transition metals.

Electronic thermal conductivity. Table 1 shows that near and above the Debye temperature, for some metals with high values of electrical resistivity, the lattice component of thermal conductivity may amount to ten percent or more of the overall thermal conductivity. In typical metals thermal energy is transported mainly by electrons, so that above θ_D, λ_e accounts for 90–99% of the value of λ.

For pure metals at high temperatures, the contribution of impurities to electronic thermal resistivity W_{e-i} follows the Wiedemann-Franz-Lorenz law according to the data of [2, 12, 23]

$$W_{e\text{-}i} = \rho_i/(L_0 T) \tag{23}$$

where ρ_i is the residual resistivity $\rho_i \approx \rho_{4.2K}$, and L_0 is the standard Lorenz number $L_0 = 2.445 \cdot 10^{-8}\ V^2/K^2$. This contribution decreases rapidly with the increase in temperature. For sufficiently pure metals at temperatures above $\frac{1}{3}\theta_D \div \theta_D$, this contribution is not noticeable against the background of other components.

The thermal resistivity of ordinary metals caused by the electron-phonon scattering has been extensively investigated in theoretical studies [2, 12]. The main and well known conclusion is that at high temperatures the W-F-L law is a good approximation for metals: $W_{e-ph} = \rho/(L_0 T)$, where ρ is the total electrical resistivity. If $\rho \approx \rho_{e-ph}$, then the electron-phonon thermal resistivity can be expressed as

Table 1 The lattice component of the thermal conductivity coefficients for several metals at $T = 2\theta_b^{\circ}$, calculated from eqs. (21, 22) using the constants, given in this study, and also in [23]

Metal	z_e	$\dfrac{W_{ph-ph}}{T} \cdot 10^1$, m/W	λ_{exp}, W/m·K	W_{ph-e}, K·m/W	λ_g,W/m·K	λ/λ_g, %
Sc	2	0.35	18	0.47	2.0	6—11
Y	2	1.66	16	0.43	2.0	9—12
La	2	2.35	15	0.39	2.2	14—16
Ti	2	0.99	19	1.08	0.9	5—10
Zr	2	0.75	22	0.84	1.1	5—9
Hf	2	0.98	23	0.46	2.0	8—10
V	2	1.25	40	0.4	2.0	5—12
Nb	1	2.04	55	0.05	6.2	10—20
Ta	2	4.1	52	0.23	2.4	6—13
Cr	1	0.8	66	0.06	8.0	12—20
Mo	1	0.64	145	0.03	13	9—12
W	2	0.64	144	0.11	7.0	6—10
Mn	2	2.2	11	0.13	3.4	31—40
Tc	2	1.6	52	0.24	2.7	5—9
Re	2	1.84	63	0.4	2.0	4—9
Fe	2	1.41	45	0.23	3.0	7—15
Co	2	1.90	57	0.06	4.9	8—15
Ni	2	2.04	61	0.1	4.2	7—17
Ru	1	1.42	100	0.36	2.1	6—10
Rh	1	1.42	128	0.03	7.8	6—10
Pd	1	2.84	80	0.05	4.8	6—10
Os	2	0.38	90	0.17	5.0	6—10
Ir	2	2.83	140	0.11	8.2	6—8
Pt	1	2.68	73	0.05	5.7	6—13
Ce	2	1.49	12	0.24	3.5	20—30
Pr	2	1.87	12	0.43	2.1	15—30
Nd	2	2.0	14	0.40	2.2	15—25
Sm	2	0.98	12.5	0.50	1.9	15—25
Eu	2	2.72	10	0.32	2.6	20—30
Gd	2	0.81	11	0.23	3.9	25—30
Tb	2	1.75	12	0.32	2.6	20—30
Dy	2	1.62	11	0.51	1.8	15—25
Ho	2	1.49	13	0.20	4.0	30—35
Er	2	1.44	14	0.31	2.8	15—20

$$W_{e-ph} = \rho_{e-ph}/(L_0 T) = A/L_0 \tag{24}$$

and it does not depend on temperature.

Using, for the first approximation, a model of a spherical Fermi surface, a simple Debye model for the phonon spectrum, and taking into account only the ordinary processes of electron-phonon scattering, Wilson obtained the following expression for thermal resistivity

$$W_{e \cdot ph} = \frac{1}{L_0 T} 4\rho\theta_D \left\{ \left(\frac{T}{\theta_D}\right)^5 \left[I_5\left(\frac{\theta_D}{T}\right) - \frac{1}{2\pi^3} I_7\left(\frac{D_D}{T}\right) \right] \right.$$

$$\left. + \frac{3}{\pi^2} \left(\frac{R_F}{D_W}\right)\left(\frac{T}{\theta_D}\right)^3 I_6\left(\frac{\theta_D}{T}\right) \right\}$$

(25)

where I^5 and I^7 are the Fermi integrals (tabulated, for example, in study [12]); R_F and D_W, are the radii of the Fermi and Wigner-Seitz spheres. In the Wigner-Seitz approximation, the ratio R_F/D_W, is related to the number of free electrons per atom $R_F/D_W = (z/2)^{2/3}$ [12]. At low temperatures $W_{e-ph} = WT^2$.

Taking into account that, at low temperatures, scattering by impurities becomes substantial, then the general expression for the low-temperature function is $W_{e-ph} = A/T + WT^2$ [2, 12].

The term WT^2 is due to the inelastic scattering of electrons. Different times of relaxation should be introduced for the processes of the dispersion of heat and charge carriers.

Electrical resistivity and electronic thermal conductivity are determined by the same carriers and types of scattering. It is, therefore, customary to analyze the behavior of the Lorenz function: $L = \lambda\rho/T$.

Several reasons exist for $L \neq L_0$. The first is the inelastic character of scattering. The scattering processes causing electrical and thermal resistivity have different intensities. It is impossible to introduce a single relaxation time for electrical and thermal resistivity. The other reasons are: the complex structure of the electronic zones and spectrum, as well as the non-smooth nature of $N(\varepsilon)$ within the boundaries of the thermal energy layer (the width of which is on the order of kT) [12].

In the transition metals, the role of band effects attributed to the 'nonsmoothness' of $N(\varepsilon)$ becomes very visible. One of the ways of taking both this factor and inter-band scattering into account is based on the momentum method [2, 12]. Similar to the case of electrical conductivity, electronic heat conductivity can be expressed as

$$\lambda_e = k^2 T \left(M_2 - \frac{M_1^2}{M_0} \right)$$

where M_n is given by eq. (12).

Similarly, for the Lorenz functions we have

$$L = (k/e)^2 (M_2/M_0) - S^2$$

where S is equal to the absolute thermoelectric power of the given metal.

For electronic thermal conductivity of the d-transition metals within the s-d model of Mott, the calculations [11] give

$$\lambda_e \approx A\theta_D^2 N(\varepsilon) \left[1 + \pi^2/6 \, (kT)^2 \left\{ 37/5N_1^2 - 21/5N_2 \right\} \right]$$

(26)

where the symbols are the same as in eq. (13). For the reduced Lorenz function we have:

$$L/L_0 = 1 + \frac{\pi^2}{6} (kT)^2 \left[\frac{37}{5} N_1^2 - \frac{21}{5} N_2 \right] / \left\{ 1 + \frac{\pi^2}{6} (kT)^2 \right.$$

$$\left. \times \left[3N_1^2 - N_2 \right] \right\} \tag{27}$$

It is obvious that, in these equations, the type of temperature dependence is defined by the shape of the curve of density of electronic states. As was mentioned before, these expressions are correct only in the vicinity of the Debye temperature. They are useful not so much for plotting the temperature dependences of ρ, λ, and S, as for the analysis of the probable role of band contributions to these dependences.

It is useful to add two more expressions related to thermoelectric power and magnetic susceptibility [11]

$$S = \frac{(\pi k)^2}{3e} N_1 (\varepsilon) T \left[1 + \frac{\pi^2}{6} (kT)^2 \left\{ \frac{37}{5} N_1^2 - \frac{42}{5} N_2 \right\} \right] \tag{28}$$

$$\chi = 2\mu_B^2 N (\varepsilon) \left[1 - \frac{\pi^2}{6} (kT)^2 \left\{ N_1^2 - N_2 \right\} \right] \tag{29}$$

where μ_B is the Bohr magneton.

The temperature dependence of the position of the Fermi level was neglected in the above expressions. Therefore, they may be used only for the narrow temperature region near and above the Debye temperature, where the terms of the expansion series are small in comparison with unity.

Nevertheless, they are very convenient for the qualitative analysis of kinetic properties. According to these expressions, the value, and more importantly, the type of temperature dependences of the discussed physical properties are one-to-one functions of the density of electronic states and its derivatives. The change in the phonon spectrum is taken into account by the temperature dependence of the Debye temperature.

Two possible approaches exist for comparing eqs. (26)–(29) with the experimental results. On the one hand, if, for some metals, the values of $N(\varepsilon)$, $\partial N(\varepsilon)/\partial \varepsilon$ and $\partial^2 N(\varepsilon)/\partial \varepsilon^2$ can be determined accurately, they may be used for the direct calculation of the values of ρ, S, λ, χ using formulas (13), (26)–(29). However, the dependence of the density of states on energy is not directly measureable, and can only be determined indirectly, on the basis of various physical properties. However this requires the introduction of certain models and correcting constants, e.g., a constant for the electron-phonon interaction. Theoretical calculations of $N(\varepsilon)$ cannot usually claim high accuracy, especially for the derivatives $N_1(\varepsilon)$ and $N_2(\varepsilon)$.

On the other hand, the eqs. (26)–(29) can be regarded as a system of equations with three unknowns: $N(\varepsilon)$, $N_1(\varepsilon)$ and $N_2(\varepsilon)$. It is possible to establish relationships among different properties using these equations. Even in zero-approximation, thermoelectric power is determined by the first derivative of $S \sim N_1(\varepsilon)$, and the second derivative terms in the equations for $\rho(T)$ and $\chi(T)$ coincide. Therefore, the values of ρ and λ can be expressed as

$$\rho = ATN\,(\varepsilon)\left[1 - \frac{3}{2}\,\frac{S^2}{L_0} + \left\{\frac{\chi}{2\mu_B\,N\,(\varepsilon)} - 1\right\}\right] = \rho_0\left[1 - \frac{\Delta\rho}{\rho_0}\right]$$

$$\lambda = \frac{1}{A}\,\frac{1}{N\,(\varepsilon)}\,L_0\left[1 + \frac{37}{10}\,\frac{S^2}{L_0} - \frac{21}{5}\left\{\frac{\chi}{2\mu_B N\,(\varepsilon)} - 1\right\}\right] = \lambda_0\left[1 + \frac{\Delta\lambda}{\lambda_0}\right]$$

(30)

Calculation of the terms in equations (26)–(29) shows that the contribution of the $N_1(\varepsilon) \sim S^2/L_0$ term is usually very small (it reaches about 10% only for palladium and platinum) (Table 2).

If the above discussed fact is taken into account, eq. (30) can be simplified, because deviations from ordinary expressions ($\rho \sim T$, $\lambda \sim T^\circ$, $\chi \sim T^\circ$) are only caused by the terms which include N_2. These terms can be expressed through $\chi(T)$

$$\rho \approx AT\,\frac{\chi\,(T)}{2\mu_B}\;,\;\lambda = \lambda_0\left\{1 - \frac{21}{5}\left[\frac{\chi\,(T)}{2\mu_0\,N\,(\varepsilon)} - 1\right]\right\}$$

(31)

It is evident from these relations that, if in metals, the value of χ decreases when the temperature increases, one can observe a deviation in $\rho(T)$ from the linear dependence ($\partial V\rho/\partial T^2 < 0$) and an increase in the total thermal conductivity. (This phenomenon occurs for the transition metals of the III, V and VIII subgroups).

For easy quantitative comparison of results, however, it is advisable, following from the Mott-Shimizu theory, to estimate the second derivatives in eqs. (26)–(29) from the temperature dependences of $\chi(T)$, $\rho(T)$ and $\lambda(T)$. It has been mentioned that first derivatives obtained from the data on thermoelectric power are small in comparison with second derivatives. Having designated by N_2^χ, N_2^ρ and N_2^λ the second derivatives obtained from the temperature dependencies of χ, ρ and λ, and using equations (26)–(29) (for example, $N_2^\rho = [\rho N(\varepsilon)AT]^{-1} - 1$), it is

Table 2 The parameter values in the expansion of kinetic coefficients by equations (13), (26)–(29), computed from the experimental dependences of $\chi(T)$, $\rho(T)$ and $\lambda(T)$ near 500 K

Metal	S_{max}^2/L_0	N_2^χ, eV^{-2}	N_2^ρ, eV^{-2}	N_2^λ, eV^{-2}	Metal	S_{max}^2/L_0	N_2^χ, eV^{-2}	N_2^ρ, eV^{-2}	N_2^λ, eV^{-2}
Sc	0.01	−88.3	−32.6	−3.6	Rh	0.0001	+24.7	+9.4	+4.0
Ti	0.004	+31.8	−33.2	−0.39	Pd	0.1	−22.60	−12.5	−5.0
V	0.01	−12.0	−13.8	−4.0	Hf	0.01	+32.6	+7.0	−1.0
Cr	0.014	+10.3	+8.8	−4.0	Ta	0.01	−33.2	−12.7	−2.6
Y	0.0016	−49.3	−27.4	−3.0	W	0.032	+15.9	+11.0	+6.0
Zr	0.004	+39.7	−13.5	−2.0	Re	0.0025	+14.6	−14.1	−5.0
Nb	0.01	−26.3	−16.8	−4.4	Os	0.01	+44.0	−1.7	+2.6
Mo	0.032	+40.2	+11.0	+3.6	Ir	0.01	+53.0	+14.6	+4.0
Tc	0.002	+1.7	−22.1	−4.0	Pt	0.1	−49.0	−10.3	−3.6
Ru	0.01	+24.7	+1.7	+4.4					

possible to estimate the relationships between second derivatives of density of states, calculated by different methods.

Table 2 lists the values of N_2^x and N_2^ρ. It follows from the analysis of this data that the signs of N_2^x and N_2^ρ coincide, except for the metals of the titanium, rhenium and osmium subgroups. However the values of N_2^x and N_2^ρ usually differ very substantially, and N_2^x is several times larger than N_2^ρ.

Table 2 also lists the results of calculations of N_2^λ. The table shows that the signs of N_2^x, N_2^ρ and N_2^λ coincide for the metals of the scandium, vanadium, and chromium subgroups, as well as for some noble metal subgroups (iridium and rhodium, palladium and platinum). This is in agreement with eqs. (30) and (31). Their values however, often differ by a factor of several times, so that $N_2^x > N_2^\rho > N_2^\lambda$.

One of the reasons for this relation is that in expressions similar to eq. (30), the interaction among the conduction electrons was not taken into account.

All this shows that, in some cases, the Mott s-d model describes satisfactorily the temperature dependences of some kinetic properties. At high temperatures, however, this model cannot completely describe all of the experimental data.

It should be noted that the interband s-d transitions strongly influence the electronic component of thermal conductivity and the Lorenz number. A strong deviation in the Lorenz number from the standard value of L_0 can be an indicator of such transitions.

The electron-electron contributions to thermal resistivity of the regular Coulomb type are similar to the electron-electron contributions to electrical resistivity at moderate and high temperatures. They are not visible against the background of other components, although L for the thermal resistance may be considerably smaller than $L_0(L \approx 0.6\,L_0)$. In the presence of magnetic inhomogeneities, however, these contributions may amount to a considerable portion of the value of W. Similar to the case for electrical resistance, the scattering by magnetic inhomogeneities may be studied using the one-band model developed by Kasuya [15, 18], where the s-electrons are scattered by the localized spins of the d-electrons, or using the Mott band model [2, 9, 12, 15], where the s-electrons are scattered in the d-band.

Study [25] considered thermal resistance in the band model framework. It showed that for the high-temperature limit $L = L_0$, and at low temperatures $W_{e-m} \sim T^2$ and $L/L_0 = T/\theta_M$, where θ_M is a constant in units of temperature.

In the Kasuya model $L = L_0$ at $T \geq T_c$, because $\rho = AT + D$, $\lambda_e = L_0/(A + DT)^{-1}$, guaranteeing an increase in thermal conductivity with temperature.

Below the Curie temperature, the Lorenz number will depend on temperature because of the inelastic contributions. Studies [26, 32] showed that

$$L = L_0[1 - x_m^2 \pi^{-2}(\rho_s^n/\rho)]$$

where $x_m = H_0/kT$, $H_0 = sg\mu_B H_e$, g is the Lande factor; H_e is the molecular field of Weiss; $\rho_s^n \approx \rho_m$, i.e., it is equal to the magnetic contribution to electrical resistivity; s is the spin at the lattice site.

Finally, in almost-magnetic transition metals the scattering by paramagnons is also inelastic in character. According to [17], the Lorenz number depends on the temperature: $L = L_0 f(\Delta)$, where Δ is a value related to the parameters of the broad s-band and the narrow d-band. At low temperatures, thermal resistivity due to paramagnon scattering changes as $W_{e-pm} \sim T$. At high temperatures the Lorenz number, related to the electron scattering by paramagnons, is less than the standard value of L_0, and approaches L_0 when $T \to \infty$:

$$L/L_{pm} \simeq 1/(1 + \Delta/T) \tag{32}$$

It should be noted that for metals in which the Fermi level is located near the maximum of the density of states, the Mott component, related to the s-d scattering, leads to an increase in the Lorenz number in comparison with L_0. The comparison of the experimental values of L with those derived from expressions (27) and (32) permits the estimation of the influence of the band and paramagnon effects for similar metals.

The behavior of thermal conductivity near the phase transformation points. In the vicinity of the points of magnetic phase transitions of the second order, such as the Curie and Neel points, two competing mechanisms determine the behavior of thermal conductivity. On the one hand, the critical fluctuations of the spin density must form a mechanism of 'slow' thermal conductivity approaching infinity at the phase transition point [27, 28, 30]:

$$\lambda = (t)^{-\alpha'}, T < T_c, \quad \lambda = (t)^{-\alpha}, \quad T > T_c$$

where α and α' are the critical values, the same as for the specific heat and temperature coefficients of electrical resistivity.

There are dielectrics for which λ has a maximum at the critical point [27]. The physical nature of this maximum is due to the fact that critical fluctuations are the heat carriers, and the thermal energy, transported by them, increases when the temperature approaches T_c. On the other hand, in the region of the magnetic disordering point, a critical dispersion of current and energy carriers should be caused by the spin fluctuations. The corresponding contribution to thermal resistivity should have a maximum.

Calculations [26, 29, 32] for electron-electron thermal resistivity of ferromagnetics showed that anomalies in thermal and electrical resistivity are closely related and that, at high temperatures (above θ_D), the Lorenz number L does not differ from L_0. The electron components of thermal conductivity should have a weak maximum with the cusp point

$$\lambda_e = L_0 T/\rho \tag{33}$$

so that $(\partial \lambda_e/\partial T) \approx (T/\rho^2)(\partial \rho/\partial T)$

Regarding the lattice component, it should be said that the number of theoretical investigations is not sufficient for a more or less definite conclusion. Calculations [30] show the presence of a minimum in lattice thermal conductivity: $\lambda_g = B(t_T^2) - \beta$, where β is close to 0.5, for a pure dielectric, and increases to 0.75 in the presence of impurities. The problem of how much the contributions to scattering compete with the additional mechanism of heat transport requires experimental data for every separate case. The same requirement is necessary to establish the correctness of relation (33), especially for antiferromagnetic disordering.

§3 DIFFUSIVITY

Thermal conductivity, λ, is related to diffusivity, a, specific heat capacity at constant pressure, c_p, and density, d, through the equivalence relation $\lambda \equiv ac_pd$.

The propagating velocity of a temperature wave in a material defines the diffusivity value [31]. Usually, this property is not used for the analysis of the electron and phonon scattering mechanisms. In dielectrics, however, it is possible to determine such an important characteristic as the mean free path of phonons Λ_{ph}

$$\lambda = Ac_V \cdot v_l \Lambda_{ph}; \quad a = Av_l \Lambda_{ph}$$

where A is a proportionality coefficient; in the case of isotropic matter $A = \frac{1}{3}$; and, v_l is the velocity of longitudinal vibrations.

In the vicinity of the point of the second order phase transformation, the diffusivity has a critical anomaly:

$$a = A(t)^\gamma \tag{34}$$

where γ is a critical indicator, $\gamma = 0.1 \div 0.5$ [26, 27]. Diffusivity minima are definitely observed at the magnetic phase transition points. Their shapes, however, have not been sufficiently investigated.

§4 ABSOLUTE THERMOELECTRIC POWER

Equation (3), used for the determination of the absolute thermoelectric power, was given earlier. If we express S through $\sigma(\varepsilon)$, then

$$S = (\pi^2/3)(k^2T/3)\{\partial\sigma(\varepsilon)/[\sigma(\varepsilon)\,\partial\varepsilon]\}_{\varepsilon=E_F} \tag{35}$$

where the value of the derivative $\partial\sigma(\varepsilon)/\partial\varepsilon$ is taken at the point where the energy is equal to the Fermi energy. The value of thermoelectric power is very sensitive to the fine details of electronic structure and the parameters of scattering. From expression (35) we have

$$S \sim \partial\Lambda_e/(\Lambda_e\,\partial\varepsilon) + \partial S_F/(S_F\,\partial\varepsilon) \tag{36}$$

where Λ_e is the mean free path of electrons; and, S_F is the Fermi surface area. It follows from expressions (35) and (36) that thermoelectric power can be positive or negative, depending not only on the sign of the charge of the principal carriers, but also on the peculiarities of the electronic structure and the scattering parameters. For free electrons

$$S = 1/3(\pi^2 k^2)e^{-1}(T/E_F) \tag{37}$$

The above expression shows that the absolute thermoelectric power should be negative and linearly dependent on temperature. For typical metals at room temperature, it should amount to several $\mu V/K$.

It is well known that even alkali metals satisfy expression (37) poorly [2], and that lithium, on the contrary, has positive values for S.

In the transition metals, the situation is even more complex. According to Mott's model of s-d scattering and equation (28), thermoelectric power is proportional to the first derivative of the density of electronic states with respect to energy: $S \approx N_1(\varepsilon)T$. If we estimate S from the dependence of $N(\varepsilon)$, then for many metals the calculated values for S may exceed the corresponding experimental values by more than one order of magnitude. Such a discrepancy in values indicates that other mechanisms of scattering are predominant, and only a small portion of carriers in these metals is scattered according to the band mechanism, accompanied by changes in the effective masses of the s-carriers because of their transfer to the d-band. In particular, the scattering by magnetic inhomogeneities in ferromagnetics was investigated in studies [29, 32, 33]. In the region of the Curie and Neel points, thermopower values have a discontinuity or a more complex anomaly.

§5 THE HALL COEFFICIENT

For the simple case of free electrons, the Hall coefficient is inversely proportional to the electron concentration

$$R = (ne)^{-1} \tag{38}$$

Two kinds of Hall coefficients can be distinguished: isothermal and adiabatic. The former coefficient has been measured more frequently in practice [2, 15]. Using an approximation for the relaxation time and the model of nearly-free electrons, the expression for the Hall coefficient can be presented [15] as: $R = e\tau(m^*\sigma)^{-1}$, where m^* is the effective mass of the carrier. The value of m^* is calculated from the second derivative of energy with respect to the wave number. Therefore, the sign of the Hall coefficient depends on the features of the Fermi surface. In this approximation, the sign of the Hall constant is negative for electrons and positive for holes [2, 15].

In the two-band model, the Hall coefficient appears as [2, 15]

$$R = e\left(\tau_1\sigma_1/m_1^* + \tau_2\sigma_2/m_2^*\right)/(\sigma_1 + \sigma_2)^2 \tag{39}$$

where σ_1 and σ_2 are the contributions to the conductivity of the first and second bands; m_1^* and m_2^*, and τ_1 and τ_2 are the effective masses of the carriers and their relaxation times, respectively.

At low temperatures, the Hall coefficient is highly affected by the fine features of the topology of the Fermi surface. It may depend substantially on the magnitude of the magnetic field (the Hall coefficient is treated separately for weak and strong magnetic fields) [15].

In ferromagnetics, the Hall electromotive force is defined as $\varepsilon_H = R_0B + R_sI$, where R_0 is the regular or normal Hall coefficient; R_s is the anomalous Hall coefficient; B is the induction from the magnetic field; I is magnetization of the sample.

At room temperature in ferromagnetics, the value of R_s is greater than R_0 by 1–2 orders of magnitude. It is strongly dependent on temperature (a correlation between the behavior of R_s and the square of the electrical resistivity is often observed [15]). The Hall coefficient data in this book will be given only for the paramagnetic metals.

§6 ELASTIC PROPERTIES

The classical theory of elasticity postulates Hooke's law, which establishes a linear relation between the tensors of stress σ_{ik} and the strain ε_{jl} [34]

$$\sigma_{ik} = \sum_{j,l} c_{ik,jl}\,\varepsilon_{jl}$$

The proportionality coefficients $c_{ik,jl}$ are called the elastic constants (EC) or the components of the elastic stiffness, and they are measured in the units of pascals. Altogether there are 81 elastic constants. The symmetry between the tensors of stress and strain, and the existence of an elastic potential decrease the number of independent EC down to 21 for the triclinic system of the lowest symmetry. To describe the elastic properties of crystals with higher symmetry, it is required to have a smaller number of independent EC: for monoclinic system— 13, rhombic—9, trigonal—7 or 6, tetragonal—6 or 7, hexagonal—5 and cubic—3.

An isotropic solid body is defined by two EC: the Lame constants, λ_l and μ_l, which are related to the macroscopic elastic moduli (EM) as follows: the shear modulus $G = \mu_l$, Young's modulus $E = \mu_l(3\lambda_l + 2\mu_l)\mu_l/(\lambda_l + \mu_l)$, and the bulk modulus $K = \lambda_l + 2\mu_l/3$.

In an isotropic solid body the longitudinal strain l_l is accompanied by l_t, the strain of the opposite sign in the transverse direction. The ratio of the two strains: the longitudinal stretch and the transverse contraction is called Poisson's coefficient: $\mu = |-l_l/l_t|$.

The characteristics of an elastic isotropic material are interrelated. The relationships between E, G, K and μ are shown in Table 3.

The experimental values of EC and EM given in this book have been mainly obtained from the dynamic measurements of the velocity of a wave spreading in a solid body.

The determination of all independent EC requires, for every system, a fixed number of measurements of the longitudinal (v_l) and transverse (v_t) wave velocities along the different crystallographic directions.

Table 3 The relationships between Young's (E), shear (G), and bulk (K) moduli, and Poisson's coefficient (μ) for an isotropic solid body

Determined Value	K, E	K, G	K, μ	E, G	E, μ	G, μ
K	—	—	—	$\dfrac{EG}{3(3G-E)}$	$\dfrac{E}{3(1-2\mu)}$	$\dfrac{2}{3}G\dfrac{1+\mu}{1-2\mu}$
E	—	$\dfrac{9KG}{3K+G}$	$3K(1-2\mu)$	—	—	$2G(1+\mu)$
G	$\dfrac{3KE}{9K-E}$	—	$\dfrac{3}{2}K\dfrac{1-2\mu}{1+\mu}$	—	$\dfrac{E}{2(1+\mu)}$	—
μ	$\dfrac{1}{2}-\dfrac{E}{6K}$	$\dfrac{1}{2}\dfrac{3K-2G}{3K+G}$	—	$\dfrac{E}{2G}-1$	—	—

Table 4 lists the relationships between the EC and the longitudinal and transverse wave velocities for different crystal structures.

The conditions of stability of a crystal lattice require the satisfaction of certain relations between the elastic constants [35]. For the cubic system, such relations are: $1/2(c_{11}-c_{12}) > 0$ and $(c_{11}+2c_{12}) > 0$.

The former expresses the lattice resistance to shear stress; the latter is equal to $\frac{1}{3}K$ (K is the bulk modulus). It follows from these two conditions that $c_{11}/2 < c_{12} < c_{11}$.

For a hexagonal crystal, the stability conditions are $(c_{11}^2 - c_{12}^2) > 0$; $c_{33}(c_{11} + c_{12}) - 2c_{13}^2 > 0$; $(c_{11}c_{33} - c_{13}^2) > 0$.

It follows from these inequalities that $-c_{11} < c_{12} < c_{11}$; the shear constant, which was first introduced by Huntington, is $c_H = (c_{11} + c_{12} + 2c_{33} - 4c_{13})$; and, the volume compressibility of a hexagonal crystal is always positive. The value of c_{13} should be less than the geometric average of c_{11} and c_{33}. The relations for the elastic constants of tetragonal and orthorhombic crystals are similar.

In addition to the EC, this book lists the data for the bulk modulus and the volume compressibility $\chi = 1/K \neq \chi = -(1/V)(dV/dP)$, which relates the relative change in volume to the bulk stress.

For the cubic crystals, volume compressibility is $\chi = 1/K = 3/(c_{11} + 2c_{12})$. The compressibility of a single crystal is equal to the compressibility of a mixture of single crystals with different orientations. Therefore, the measurements of single crystal compressibility give the compressibility value for a polycrystal [34].

For hexagonal and tetragonal crystals compressibility is

$$\chi = (c_{11} + c_{12} + 2c_{33} - 4c_{13})/[c_{33}(c_{11} + c_{12}) - 2c_{13}^2]$$

Table 4 The relationships between the measured velocity of sound and the elastic constants in crystals of different symmetries

Direction of Propagation	Direction of Polarization	Mode of Vibration	Relationships Between the Velocities and Elastic Constants
		Tetragonal System	
[100]	[100]	L	$dv^2 = c_{11}$
[100]	[001]	T	$dv^2 = c_{44}$
[100]	[010]	T	$dv^2 = c_{66}$
[001]	[001]	L	$dv^2 = c_{33}$
[001]	[100]	T	$dv^2 = c_{44}$
[110]	[110]	L	$dv^2 = 1/2\,(c_{11} + c_{12}) + c_{66}$
[110]	[00$\bar{1}$]	T	$dv^2 = c_{44}$
[110]	[1$\bar{1}$0]	T	$dv^2 = 1/2\,(c_{11} - c_{12})$
Under Angle 45° к [010]	[011]	QT T	$dv^2 = 1/4\,(c_{11} + c_{33} - 2c_{13})$ $dv^2 = 1/2\,(c_{44} + c_{66})$
и [001]	[100]	QL	$dv^2 = 1/4\,(c_{11} + c_{33} + 2c_{13} + 4c_{44})$
		Hexagonal System	
[100]	[100]	L	$dv^2 = c_{11}$
[100]	[001]	T	$dv^2 = c_{44}$
[100]	[010]	T	$dv^2 = 1/2\,(c_{11} - c_{12})$
[001]	[001]	L	$dv^2 = c_{33}$
[001]	$\perp$ [001]	T	$dv^2 = c_{44}$
[101] [101]	[101] $\perp$ [101]	QL QT	$dv^2 = 1/4\,(c_{11} + c_{33} + 2c_{44}) \pm$ $\pm\, 1/2\,[^1/_4(c_{11} - c_{33})^2 + (c_{13} + c_{44})^2]^{1/2}$
[101]	[010]	T	$dv^2 = 1/4\,(c_{11} - c_{12} + 2c_{44})$
		Cubic System	
[001]	[001]	L	$dv^2 = c_{11}$
[001]	$\perp$ [001]	T	$dv^2 = c_{44}$
[110]	[110]	L	$dv^2 = 1/2\,(c_{11} + c_{12} + 2c_{44})$
[110]	[1$\bar{1}$0]	T	$dv^2 = 1/2\,(c_{11} - c_{12})$
[110]	[00$\bar{1}$]	T	$dv^2 = c_{44}$
[111]	[111]	L	$dv^2 = 1/3\,(c_{11} + 2c_{12} + 4c_{44})$
[111]	$\perp$ [111]	T	$dv^2 = 1/3\,(c_{11} - c_{12} + c_{44})$

L, T, QL, QT are longitudinal, transverse, quasilongitudinal, and quasitransverse modes of vibration, respectively.

The elastic constants are defined as second derivatives of the density of energy W with respect to strain

$$c_{ij} = \partial T_i/(\partial e_j) = \partial^2 w/(\partial e_j \partial e_i) = \partial^2 w/(\partial e_i \partial e_j).$$

Therefore, the elastic constants are related to other physical properties which participate in the expressions for the energy of a solid body, especially those which are the subjects of study in lattice dynamics.

The Debye temperature, which is determined from the elastic constants θ_D', is often used. When the value of θ_D is determined from the elastic constants, the Debye theory establishes the contribution of lattice vibrations to the specific heat capacity with an accuracy of only 10–20% for the major part of the investigated temperature interval. This happens because the Debye theory considers a solid body as an elastic continuum in which all waves travel with the same velocity, independent of their wavelength. Such a model is satisfactory only for the limiting cases of long wavelengths or low temperatures. Therefore, the calculation of an accurate value for θ_D from the elastic constants is practically possible only for the low temperatures [36].

The calculation of θ_D from the elastic constants is performed using the formula from [36]

$$\theta_D = h/k \, (3pN/4\pi M)^{1/3} \, v_m$$

where h and k are Plank and Boltzmann constants, respectively; N is the Avogadro number; M is the molecular mass of a substance; p is the number of atoms in a molecule; v_s is the average velocity of sound. For isotropic crystals

$$v_m = [{}^1\!/_3(2/v_t^3 + 1/v_t^3)]^{-\frac{1}{3}}$$

Study [36] presented the methods for calculating θ_D^E for crystals with cubic to orthorhombic symmetries.

This book summarizes the data on the temperature dependence of the elastic constants and moduli over a broad temperature interval. Theory [34] predicts the following behavior for the elastic moduli: approaching $T = 0$ K with a zero slope,

Table 5 The values of the coefficient $A = c_{44}/c_{66}$ for metals with HCP structure [38]

Metal	$A_{4.2\ K}$	$A_{300\ K}$	$A_{923\ K}/A_{1128\ K}$	Metal	$A_{4.2\ K}$	$A_{300\ K}$	$A_{923\ K}/A_{1128\ K}$
Cd	0.542	0.536	—	Co	1.07	1.07	—
Zn	0.660	0.610	—	Hf	1.034	1.072	—
Zr	0.829	0.907	1.480/2.00	Ti	1.139	1.327	2.103/2.776
Re	0.917	0.937	0.974/0.976	Tl	2.581	2.688	—
Ru	0.972	0.964	0.923/—	Gd	1.039	0.995	—
Mg	0.981	0.971	—	Lu	0.984	0.989	—
Dy	—	1.000	1.126/—	Ho	1.004	1.020	—
Er	0.928	1.007	—	Nd	0.956	0.970	—
Y	1.009	1.018	1.129/—				

Table 6 The values of several physical constants and temperatures of the phase transitions in metals [39, 40, 74, 77, 82, 100, 227, 321]

Metal	Atomic number	Atomic mass	Density, g/cm³	Crystal structure	T_{struc}, K	$T_{supercond}$	T_C	T_N	T_m	Debye temperature, K $T = 0$ K; $\dfrac{\theta^0_{calor}}{\theta^0_{elast}}$	$T = 298$ K; $\dfrac{\theta^0_{calor}}{\theta^0_{elast}}$
1	2	3	4	5	6	7	8	9	10	11	12
Li	3	6.939	0.534	Distort. bcc (α), bcc (β)	75 (α–β)	—	—	—	453.7	$\dfrac{352}{—}$	$\dfrac{448}{—}$
Be	4	9.012	1.85	Hcp (α), bcc (β)	1533 (α–β)	6	—	—	1550	$\dfrac{1160}{1462}$	$\dfrac{1031}{—}$
Na	11	22.990	0.971	Distort. bcc (α), bcc (β)	35 (α–β)	—	—	—	371	$\dfrac{157}{—}$	$\dfrac{155}{—}$
Mg	12	24.312	1.74	Hcp	—	—	—	—	923	$\dfrac{396}{388}$	$\dfrac{330}{—}$
Al	13	26.982	2.702	Fcc	—	1.196	—	—	933.52	$\dfrac{423}{430}$	$\dfrac{390}{—}$
K	19	39.102	0.86	Fcc	—	—	—	—	336.8	$\dfrac{89}{—}$	$\dfrac{100}{—}$
Ca	20	40.08	1.55	Fcc (α), bcc (β)	737 (α–β)	—	—	—	1123	$\dfrac{234}{—}$	$\dfrac{230}{—}$
Sc	21	44.956	3.00	Hcp (α), bcc (β)	1607 (α–β)	—	—	—	1812	$\dfrac{490}{—}$	$\dfrac{476}{—}$
Ti	22	47.90	4.5	Hcp (α), bcc (β)	1155 (α–β)	0.39	—	—	1941	$\dfrac{426}{—}$	$\dfrac{380}{—}$

25

Metal	Atomic number	Atomic mass	Density, g/cm³	Crystal structure	T_{struc}, K	$T_{supercond}$	T_C	T_N	T_m	Debye temperature, K	
										$T=0$ K; $\dfrac{\theta^0_{calor}}{\theta^0_{elast}}$	$T=298$ K; $\dfrac{\theta^0_{calor}}{\theta^0_{elast}}$
1	2	3	4	5	6	7	8	9	10	11	12
V	23	50.942	6.1	bcc	—	5.03	—	—	2206	$\dfrac{326}{338}$	$\dfrac{390}{—}$
Cr	24	51.996	7.16	bcc	—	$T_{magn}=123$	—	311	2163	$\dfrac{598}{591}$	$\dfrac{424}{—}$
Mn	25	54.938	7.43 (α) 7.29 (β) 7.18 (γ)	Cub. (α), Cub. (β), fcc (γ), bcc (δ)	1000 (α—β) 1360 (β—γ) 1410 (γ—δ)	—	—	95 (α) 580 (β) 660 (γ)	1517	$\dfrac{418}{485}$ (α)	$\dfrac{363}{473}$ (α)
Fe	26	55.847	7.87	Bcc ferro (α), Bcc para (β), fcc (γ), bcc (δ)	1183 (β—γ) 1667 (γ—δ) —	—	1042	—	1810	$\dfrac{457}{478}$	$\dfrac{373}{—}$
Co	27	58.933	8.862	hcp (α), fcc (β)	700 (α—β)	—	1394 (α) 1130 (β)	—	1767	$\dfrac{452}{469}$	$\dfrac{386}{—}$
Ni	28	58.71	8.90	fcc	—	—	629.6	—	1728	$\dfrac{476}{472}$	$\dfrac{345}{—}$
Cu	29	63.54	8.933	fcc	—	—	—	—	1357.6	$\dfrac{342}{344}$	$\dfrac{310}{—}$
Zn	30	65.37	7.140	hcp	—	0.875	—	—	692.73	$\dfrac{316}{327}$	$\dfrac{237}{—}$

Symbol	No.	At. wt.	Density	Structure	Transition temp.				B.p.				
Ga	31	69.72	5.91	Orthorhomb. (α), tetr (β)	275.6 (α—β)	1.09	—	—	302.94	317	—	240	—
As	33	74.92	5.73	Rhomb.	—	—	—	—	1090	236	—	275	—
Rb	37	85.47	1.53	bcc	—	—	—	—	312.04	54	—	59	—
Sr	38	87.62	2.60	fcc (α), hcp (β), bcc (γ)	488 (α—β) 878 (β—γ)	—	—	—	1042	147	—	148	—
Y	39	88.905	4.47	hcp (α), bcc (β)	1753 (α—β)	—	—	—	1798	268	—	214	—
Zr	40	91.22	6.57	hcp (α), bcc (β)	1135 (α—β)	0.55	—	—	—	282	—	250	—
Nb	41	92.906	8.57	bcc		9.1	—	—	2742	241	—	260	—
Mo	42	95.94	10.24	bcc	—	0.95	—	—	2901	459	—	377	—
Tc	43	98.906	11.50	hcp	—	11.2	—	—	2473	351	—	422	—
Ru	44	101.07	12.2	hcp	—	0.48	—	—	2607	500	—	415	—
Rh	45	102.905	12.45	fcc	—	—	—	—	2236	480	—	350	—
Pd	46	106.4	12.02	fcc	—	—	—	—	1827	283	275	275	—

Metal	Atomic number	Atomic mass	Density, g/cm³	Crystal structure	T_{struc}, K	Temperatures of the phase transformations, K				Debye temperature, K	
						$T_{supercond}$	T_C	T_N	T_m	$T=0$ K; $\dfrac{\theta^0_{calor}}{\theta^0_{elast}}$	$T=298$ K; $\dfrac{\theta^0_{calor}}{\theta^0_{elast}}$
1	2	3	4	5	6	7	8	9	10	11	12
Ag	47	107.870	10.5	fcc	—	—	—	—	1235.08	$\dfrac{228}{226}$	$\dfrac{221}{-}$
Cd	48	112.40	8.65	hcp	—	0.519	—	—	594.26	$\dfrac{252}{213}$	$\dfrac{321}{-}$
In	49	114.82	7.3	tetr	—	3.40	—	—	429.78	$\dfrac{108}{111}$	$\dfrac{129}{-}$
Sn	50	118.69	5.75 (α)	fcc (α),	286.2 (α—β)	3.722	—	—	505.12	$\dfrac{236}{-}$ (α)	$\dfrac{254}{-}$ (α)
			7.31 (β)	tetr (β)						$\dfrac{196}{201}$ (β)	$\dfrac{170}{-}$ (β)
Sb	51	121.75	6.684	Rhombohed.	—	—	—	—	903.5	$\dfrac{150}{-}$	$\dfrac{200}{-}$
Cs	55	132.905	1.873	bcc	—	—	—	—	301.55	$\dfrac{40}{-}$	$\dfrac{43}{-}$
Ba	56	137.34	3.5	bcc	—	—	—	—	902	$\dfrac{110.5}{-}$	$\dfrac{116}{109}$
La	57	138.92	6.18	Double hcp (α), fcc (β), bcc (γ)	583 (α—β) 1141 (β—γ)	4.9 (α) 6.3 (β)	—	—	1193	$\dfrac{142}{-}$	$\dfrac{135}{-}$

№		Atomic weight	Density	Structure		T_{magn}						
58	Ce	140.12	6.90	fcc (α), double hcp (β), fcc (γ), bcc (δ)	143 (α—β) 348 (β—γ) 983 (γ—δ)	1.7 at 50 kbar	—	—	12.8 (β)	1077	$\frac{146}{139}$ (β)	$\frac{138}{-}$
59	Pr	140.907	6.769	Double hcp (α), bcc (β)	1069 (α—β)	—	—	—	25	1192	$\frac{85}{146}$	$\frac{138}{146}$
60	Nd	144.24	7.007	Double hcp (α), bcc (β)	1128 (α—β)	$T_{magn}=7.3$	—	—	19.55	1292	$\frac{159}{163}$	$\frac{148}{154}$
62	Sm	150.35	7.54	Hex (α), bcc (β)	1197 (α—β)	—	13.3	105	1345	$\frac{116}{169}$	$\frac{184}{165}$	
63	Eu	151.96	5.245	bcc	—	—	—	90	1099	$\frac{127}{-}$	—	
64	Gd	157.25	7.87	Hcp (α), bcc (β)	1535 (α—β)	$T_{magn}=228$	291.8	—	1585	$\frac{170}{184}$	$\frac{155}{172}$	
65	Tb	158.924	8.25	Hcp (α), bcc (β)	1560 (α—β)	—	219	227.5	1630	$\frac{150}{165}$	$\frac{158}{174}$	
66	Dy	162.50	8.556	Hcp (α), bcc (β)	1657 (α—β)	—	83.5	177.5	1682	$\frac{172}{183}$	$\frac{158}{180}$	
67	Ho	164.93	8.80	Hcp (α), bcc (β)	1701 (α—β)	—	17.5	131.6	1743	$\frac{114}{188}$	$\frac{161}{183}$	
68	Er	167.26	9.06	Hcp (α), bcc (β)	1643 (α—β)	—	19.9	84	1778	$\frac{134}{193}$	$\frac{163}{190}$	
69	Tm	168.93	9.32	Hcp (α), bcc (β)	$T_{\alpha-\beta} \leqslant T_m$	—	22	55	1818	$\frac{127}{200}$	$\frac{167}{191}$	
70	Yb	173.04	7.02	Fcc (α), bcc (β)	1065 (α—β)	—	—	—	1097	$\frac{118}{117}$	$\frac{-}{113,6}$	

Metal	Atomic number	Atomic mass	Density, g/cm³	Crystal structure	T_{struc}, K	Temperatures of the phase transformations, K				Debye temperature, K	
						$T_{supercond}$	T_C	T_N	T_m	$T=0$ K; $\dfrac{\theta^0_{calor}}{\theta^0_{elast}}$	$T=298$ K; $\dfrac{\theta^0_{calor}}{\theta^0_{elast}}$
1	2	3	4	5	6	7	8	9	10	11	12
Lu	71	174.97	9.85	Hcp (α), bcc (β)	$T_{\alpha-\beta} \lesssim T_m$	—	—	—	1923	$\dfrac{210}{-}$	$\dfrac{116}{-}$
Hf	72	178.49	13.28	Hcp (α), bcc (β)	2030 (α—β)	0.35	—	—	2220	$\dfrac{256}{-}$	$\dfrac{213}{-}$
Ta	73	180.95	16.6	Bcc	—	4.48	—	—	3269	$\dfrac{247}{-}$	$\dfrac{225}{-}$
W	74	183.85	19.3	Bcc	—	0.05	—	—	3693	$\dfrac{388}{384}$	$\dfrac{312}{-}$
Re	75	186.20	21.1	Hcp	—	1.70	—	—	3463	$\dfrac{429}{-}$	$\dfrac{275}{-}$
Os	76	190.2	22.48	Hcp	—	0.71	—	—	3283	$\dfrac{500}{-}$	$\dfrac{400}{-}$
Ir	77	192.2	22.5	Fcc	—	1.03	—	—	2723	$\dfrac{425}{-}$	$\dfrac{228}{-}$
Pt	78	195.09	21.45	Fcc	—	—	—	—	2045	$\dfrac{234}{-}$	$\dfrac{225}{-}$
Au	79	196.97	19.3	Fcc	—	—	—	—	1337.4	$\dfrac{165}{-}$	$\dfrac{178}{-}$

Element	Z	At. weight	Density	Structure	Transition (°C/K)				m.p.		
Hg	80	200.59	13.55	Bc tetr (α), rhomb (β)	79 (α—β)	3.95 (α) 4.15 (β)	—	—	234.89	$\frac{76}{-}$	$\frac{92}{-}$
Tl	81	204.37	11.85	Hcp (α), bcc (β)	508 (α—β)	—	—	—	576.2	$\frac{88}{-}$	$\frac{91}{-}$
Pb	82	207.19	11.34	Fcc	—	7.19	—	—	600.43	$\frac{102}{-}$	$\frac{87}{-}$
Bi	83	208.98	9.78	Rhomb.	—	—	—	—	544.59	$\frac{119}{-}$	$\frac{116}{-}$
Th	90	232.04	11.7	Fcc (α), bcc (β)	1673 (α—β)	1.37	—	—	2023	$\frac{170}{163}$	$\frac{100}{-}$
Pa	91	(231)	15.37	Bc. tetr. (α), bcc (β)	1473 (α—β)	1.4	—	—	1850	$\frac{159}{-}$	$\frac{262}{-}$
U	92	238.03	19.07	Orthorhomb. (α), tetr. (β), Bcc (γ)	23 (α'—α$_0$) 37 (α$_0$—α) 941 (α—β) 1048 (β—γ)	0.2—1.7	—	—	1405	$\frac{200}{246}$ (α)	$\frac{300}{244}$ (α)
Np	93	237.048	20.46	Orthorhomb. (α), tetr (β), bcc (γ)	554 (α—β) 850 (β—γ)		—	—	913.2	$\frac{121}{-}$	$\frac{163}{-}$
Pu	94	(242)	19.74	Monocl. (α), Bc monoc (β), fc. orthorhomb (γ), fcc (δ), fc tetr (δ'), bcc (ε)	396 (α—β) 475 (β—γ) 597 (γ—δ) 729 (δ—δ') 757 (δ'—ε)		—	60	912.7	$\frac{171}{-}$	$\frac{176}{-}$
Am	95	(243)	11.7	Double hcp	—	—	—	—	1473	—	—

a negative temperature coefficient at high temperatures, and a linear behavior at $T \geqslant \theta_D$.

The elastic properties of metals are determined mostly by the specific features of their magnetic and electronic structures [37].

The measurements of the elastic constants of hexagonal crystals helped to establish an empirical relationship between the shear elastic anisotropy of the hcp crystals and the hcp-bcc transformations, which occur in most of the 23 hcp metals [38]. The main criterion for the hcp-bcc transformation is an increase in the shear anisotropy coefficient $A = c_{44}/c_{66}$ with the increase in temperature prior to the phase transformation, and the condition $A > 1$ near the transformation point. This is illustrated by the data in Table 5 (gadolinium is the only exception which does not follow the above mentioned relation). The correlation between the phase transformation and the behavior of the elastic anisotropy coefficient is explained by a change in the internal crystal stress. The change in stress creates the motion of dislocations, a necessary step in starting a transformation. A change in structure starts either in the rows of the edge shear dislocations, causing an equalization of the lattice parameters, or in the packing faults connected by dislocations [38]. Table 6 lists some of the physical constants of metals.

NONTRANSITION METALS

§1 LITHIUM, SODIUM, POTASSIUM, RUBIDIUM, CESIUM, FRANCIUM

The alkali metals have low melting temperatures which decrease from 453.7 for lithium to 300.2 K for francium [39, 40]. The latter metal does not have any stable isotopes (the half-life decay period of the most long-lived, Fr^{223}, is 21 minutes and its properties are poorly known). At room temperatures, lithium, sodium and potassium have the bcc structure; rubidium and cesium the hcp structure. The parameters and temperatures of transformations are listed in Tables 6 and 7. Lithium and sodium transform into the hcp phases at 78 and 35 K, respectively. These transitions are of a martensite type, and it is very difficult to obtain a sample containing a single phase of a low-temperature modification.

Fermi surfaces of cubic alkali metals differ only slightly from the free electron sphere of the first Brillouin zone. The parameters of the Fermi surface for these metals are given in Table 7 [41]. The information regarding the elastic constants of the alkali metals of the 99.93–99.99% purity is given in Table 8. The data for lithium is given for the temperature above the α-β transition point, even though study [42] states that the temperature dependences of the elastic constants and the compressibility coefficients do not show the essential anomaly in the region below 78 K.

The elastic constants of sodium decrease linearly with the increase in temperature from 80 to 370 K. The information on elastic constants of sodium for the temperature interval 80–300 K is given in Table 8 [43]. For temperatures above 300 K in the solid state, it is described by the expression $c_{11} = (7.81-$

Table 7 Parameters of Fermi surface of alkali metals [41].

Parameter	Li	Na	K	Rb	Cs
a	6.651	8.109	10.049	10.742	11.458
m_t/m_0	1.64	1.00	1.07	1.18	1.75
m_a/m_0	1.45	1.00	1.02	1.06	1.29
A/A_0	1.06	1.00	1.03	1.06	1.12
$k_{<110>}/k_F$	1.023	1.00	1.007	1.018	1.08
$k_{<100>}/k_F$	0.973	1.00	0.994	0.980	0.94
$k_{<111>}/k_F$	0.983	1.00	0.994	0.980	0.94

Note: a - lattice parameter in atomic units; m_t - thermal mass; m_0 - mass of a free electron; m_a - optical mass; A - area of extreme cross section; k_i - radii of the Fermi surface along corresponding directions; A_0 - corresponding area for the spherical Fermi surface with the radius k_i

$0.0048T)10^9$; $c_{12} = (6.36–0.0024T)10^9$; $c_{44} = (4.59–0.0084T)10^9$ where c_{ij} is in units of Pascals (Pa). Uncertainty in the above given values is about 1% [44].

The information on the elastic constants of potassium below 195 K is listed in Table 8 [45]. Above 293 K, and almost up to the melting point, the elastic constants depend linearly on the temperature. They are described by the equations

$$c_{11} = [(3.77–0.0042T) \pm 0.2] \cdot 10^9$$

$$c_{12} = [(3.23–0.0023T) \pm 0.8] \cdot 10^9$$

$$c_{44} = [(1.98–0.0047T) \pm 0.2] \cdot 10^9$$

$$c' = [(0.27–0.0010T) \pm 0.5] \cdot 10^9 \quad [46]$$

The elastic constants of rubidium and cesium change monotonically with temperature; the corresponding information is given in Table 8. The data on the electrical resistance of solid and liquid alkali metals are compiled in [40]. The uncertainties in the values this book recommends are usually estimated as less than 5%.

From the Debye temperature to the melting point, the temperature dependence of the electrical resistivity of lithium in the solid state is close to the classical dependence $\rho = AT$ (Fig. 1). At lower temperatures, $\partial^2\rho/dT^2 > 0$. Below 80 K the data require refining because of the uncertainty in the phase composition resulting from a very slow martensite α-β transition. At the melting point the ratio of resistivities is $\rho_{liq}/\rho_{sol} = 1.59$. It should be noted that the data given above apply to a metal with a relative residual resistivity of $r = 1280$. In the liquid state, the dependence $\rho(T)$ is approximately linear, although $\partial^2\rho/dT^2 < 0$.

The temperature dependence of the electrical resistivity of sodium (Fig. 2) does not have any segments with negative curvature. The linear equation $\rho = AT$ poorly describes the experimental data. Below 35 K the experimental data require further investigation because of a possible uncertainty in the phase composition due to the martensitic transformation. Theoretical attempts to calculate the resistivities of sodium and other alkali metals have been numerous. One of the most detailed calculations was published in [47]. In this study the calculations were made with the minimum number of experimentally determined parameters (only density and one of the elastic moduli were used). The fine structure of the nonequilibrium anisotropic distribution function was taken into account. Figure 2 shows the calculated temperature dependence of the electrical resistivity of sodium. The experimental data agree with calculations to within 10–15%.

The temperature dependence of the electrical resistivity of potassium is similar

Table 8 The elastic constants of alkali metals, Pa

T, K	c_{11}	c_{12}	c_{44}	$\frac{1}{2}(c_{11}-c_{12})$
	Elastic Constants of Potassium $c_{ij} \cdot 10^{-9}$ [45]			
4.2	4.16	3.41	2.86	—
78	4.15	3.46	2.58	—
120	4.06	3.40	2.44	—
160	3.98	3.34	2.29	—
195	3.89	3.26	2.17	—
	Elastic Constants of Sodium ($c_{ij} \cdot 10^{-9}$) [43]			
80	8.57	—	5.87	0.728
140	8.32	—	5.51	0.698
200	8.11	—	5.07	0.665
260	7.86	—	4.62	0.633
300	7.69	—	4.31	0.611
	Elastic Constants of Lithium ($c_{ij} \cdot 10^{-10}$) [42]			
78	1.423	1.211	1.094	—
110	1.409	1.203	1.063	—
150	1.391	1.187	1.026	—
190	1.372	1.172	0.989	—
298	1.323	1.130	0.889	—
	Elastic Constants of Rubidium ($c_{ij} \cdot 10^{-9}$) [56]			
4.2	3.42	2.88	2.21	—
78	3.25	2.73	1.98	—
110	3.14	2.64	1.89	—
140	3.06	2.57	1.80	—
170	2.96	2.50	1.71	—
	Elastic Constants of Cesium ($c_{ij} \cdot 10^{-9}$) [57]			
4.2	2.50	2.47	1.60	—
63	2.48	2.06	1.51	—
78	2.47	2.06	1.48	—

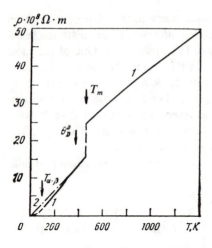

Figure 1 The temperature dependence of the specific electrical resistivity for lithium (ρ) *1*) [40]; *2*) $\rho = AT$; $A = 3.66 \cdot 10^{-10}$ $\Omega \cdot m/K$.

to that of sodium (Fig. 3). The average data in [40] refer to a metal with $r = 8500$. For this case the data cannot be described, even partially, by the relationship $\rho = AT$ because the temperature coefficient of resistivity increases with temperature. The calculations in [47] also describe the temperature dependence of $\rho\,(T)$ adequately, and the difference between the values given in [40] and [47] does not exceed 20%.

The electrical resistivity of rubidium, as a function of temperature, is also characterized by a positive curvature for the solid and liquid states over a broad temperature interval (Fig. 4). The data given in [40] apply to a metal with $r = 980$.

Cesium also has a positive curvature in the temperature dependence of the electrical resistivity for the solid and liquid states (Fig. 5). It can be deduced that

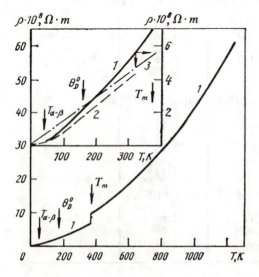

Figure 2 The electrical resistivity of sodium (ρ) vs. temperature: *1*) [40]; *2*) the calculation [47]; *3*) $\rho = AT$; $A = 1.6 \cdot 10^{-10}$ $\Omega \cdot m/K$.

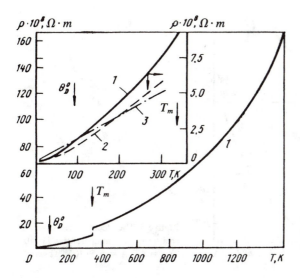

Figure 3 The electrical resistivity of potassium (ρ) vs. temperature: *1*) the data of [40]; *2*) the calculation of [47]; *3*) $\rho = AT$; $A = 1.6 \cdot 10^{-10}\ \Omega \cdot m/K$.

a similar temperature dependence of resistivity is typical for ordinary metals. The principal reasons for the increase in the slope of resistivity above the Debye temperature are thermal expansion and changes in parameters of the phonon spectrum.

The electrical resistivity of francium has not been measured. However, according to the estimate in [40], it is $\rho = 32.6 \cdot 10^{-8}\ \Omega \cdot m/K$ at 293 K.

Table 9 lists the diffusivity of alkali metals of 99.9–99.99% purity. The uncertainty of the values is $\approx 10\%$ [39]. A characteristic feature of the diffusivity is that it decreases rapidly with increases in temperature.

The thermal conductivity of lithium is shown in Fig. 6. In the solid state,

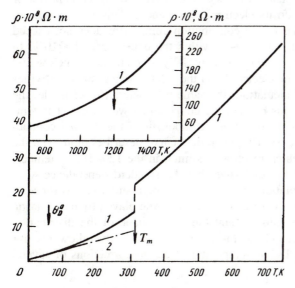

Figure 4 The electrical resistivity of rubidium (ρ) vs. temperature: *1*) [40]; *2*) $\rho = AT$; $A = 3 \cdot 10^{-10}\ \Omega \cdot m/K$.

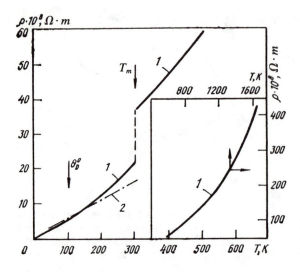

Figure 5 The electrical resistivity of cesium (ρ) vs. temperature: *1*) [40]; *2*) $\rho = AT$; $A = 5.5 \cdot 10^{-10}\ \Omega \cdot m/K$.

above 30 K, thermal conductivity decreases rapidly, mainly because of the electronic contribution. The estimation of the lattice contribution made in [23] shows that $\lambda_g/\lambda \approx 8\%$ at 300 K. The calculation of λ_e according to the W–F–L law leads to a nontrivial overestimation of λ, compared not only to λ-λ_g, but also to λ. This discrepancy is caused by the inelastic scattering of the electrons, and by different times of relaxation for the thermal and electrical resistivities. For the liquid state, within an error limit of $\approx 10\%$, total thermal conductivity is equal to the electronic component $\lambda_e^L = L_0 T/\rho$.

Figure 7 shows the temperature dependence of the thermal conductivity for sodium in the solid and liquid states. In this case as well, the overall behavior of thermal conductivity depends on its electronic component. This is because, according to the data in [23], the lattice contribution component does not exceed several percent. For the liquid state $\lambda_e \approx \lambda$ within the uncertainty of $\approx 10\%$. In the solid state λ_e becomes more noticeable, especially at temperatures below 200 K. Study [47] calculated both the electrical and the electronic thermal resistivities of sodium. For both of these calculations the same parameters (for example, the lattice parameter and the modulus c_{44}) and pseudopotential were used. The temperature dependence graph for the electronic component of the thermal conductivity is shown in Fig. 7 (curve *3*). The theoretical curve describes quite well the experimental values of λ_e^Le, which reach a maximum in the 120–180 K temperature interval. It is important to emphasize that the standard dependence $\lambda_e^g = L_0 T/\rho$ causes a substantial deviation of λ from the experimental values for temperatures below 150 K. This is evidence for the large role played by the inelastic contributions. Study [47] took into account the 'true' shape of the distribution function; this is especially evident for the thermal conductivity (allowance for electrical resistivity above the Debye temperature leads to corrections of several percent).

Table 9 The electrical resistivity (ρ) and diffusivity (a) of the alkali metals [39].

T, K	ρ·10⁸, Ω·m						a·10⁶, m²/s				
	Li	Na	K	Rb	Ce	Fr**	Li	Na	K	Rb**	Ce
50	0.162*	0.300	0.689	1.58	2.54		713	232**	229**	144	122
100	1.73	1.158	1.79	3.36	5.28	8.6	104	139**	188**	125	103
200	5.71	2.89	4.26	7.49	12.22	18.0	55.3	129**	172**	121	92.6
300	9.55	4.93	7.47	13.32	21.04	34.0	45.4	118**	157**	106	79.4
400	13.40	10.50	17.18	29.51	47.45	71	38.2	68.8	80	59.5	46.8
500	26.33	14.36	22.91	38.27	58.46	86	20.3	68.4	77.5	57.2	48.1
600	29.34	18.56	29.58	47.61	70.30	102	22.7	67.6	74.8	54.8	48.6
700	32.10	23.20	37.31	57.48	82.97	119	25.0	66.3	71.5	52.2	48.4
800	34.71	28.38	46.20	68.50	96.97	138	27.1	64.6	68.0	49.6	47.4
900	37.22	34.19	56.36	81.50	113.4	158	29.2	62.4	64.2**	47.0	46.0
1000	39.69	40.73	67.94	97.26	133.4	181	31.1	59.7	60.3**	44.2	44.0
1200	44.61	56.45	96.04	140.8	189.0	251	34.7	53.8**	—	38.6	38.7
1400	49.97	76.44	136.3	206.3	276.3	385	38.0	47.8**	—	32.5	31.9
1600	56.34*	101.8*	201.4*	301.8*	415.5*	—	41.2	—	—	25.3	25.0
1800	64.12*	135.1*	313.8*	438.2*	638.8*	—	—	—	—	16.9	17.9
2000	73.73*	184.4*	575.3*	629.4*	1000.0*	—	—	—	—	6.48	7.98
2200	85.59*	—	—	—	—	—	—	—	—	—	—

*The data require further investigation.
**Calculated values.

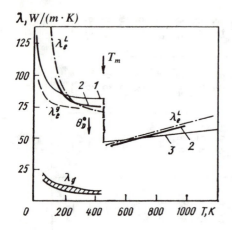

Figure 6 The effect of temperature on the thermal conductivity of lithium (λ) according to the data from: *1*) [51]; *2*) [48]; *3*) [49]; λ_g is estimate of [23].

Figure 8 shows the temperature dependence of the thermal conductivity of potassium. In the liquid state $\lambda_e \approx \lambda$ within $\pm10\%$, but, in the solid state, the difference between λ_e and λ is again apparent, especially below 200 K. For potassium as well, the contribution of the thermal lattice conductivity does not exceed 1–2% at 300 K, i.e. the thermal conductivity is essentially of the electronic nature. The data from theoretical calculations in [47] agree well with the experimental data. As in sodium, the fine structure of the nonequilibrium distribution function of potassium strongly influences the thermal conductivity.

Information on the thermal conductivity of rubidium and cesium is limited, especially for the low temperatures. Figures 9 and 10 show the data compiled in [48] and [49]. It can be claimed that the thermal conductivity has an almost exclusive electronic nature (according to estimations made in [23], the phonon contribution, at temperatures close to the Debye temperature, does not exceed 1–2%).

Figure 11 shows the temperature dependence of the Lorenz functions for the

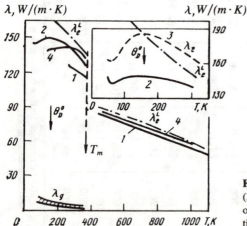

Figure 7 The thermal conductivity of sodium (λ) vs. temperature: *1*) [51]; *2*) [53]; *3*) the theoretical calculations of [47]; *4*) [48]; λ_g is estimate of [23].

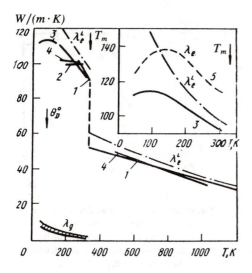

Figure 8 The thermal conductivity of potassium (λ) vs. temperature: *1*) [49]; *2*) [51]; *3* [53]; *4*) [48].

alkali metals. According to this dependence, the deviation from the standard values becomes pronounced at temperatures below 100 K. Below 20 K this difference may reach more than one order of magnitude.

The temperature dependence of the absolute thermoelectric power for all of the alkali metals, except lithium, is characterized by a monotonical growth into the negative region (Fig. 12). The thermoelectric power of lithium is positive and has a slightly pronounced maximum at room temperature.

The Hall coefficient of the alkali metals at room temperature, R_0, and the ratio R_{liq}/R_0 at the melting point are given below [113].

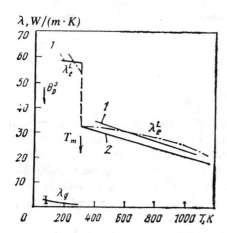

Figure 9 The thermal conductivity of rubidium (λ) vs. temperature: *1*) [48]; *2*) [49]; λ_g is estimate in [23].

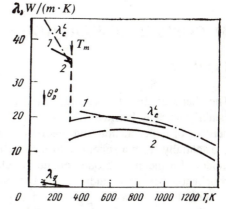

Figure 10 The thermal conductivity of cesium (λ) vs. temperature: *1*) [48]; *2*) [49].

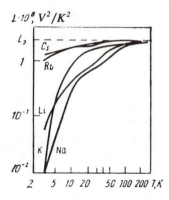

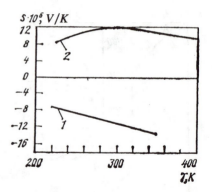

Figure 11 The Lorenz function (L) as a function of temperature for the alkali metals [54].

Figure 12 The temperature dependence of the absolute thermoelectric power (S) for the alkali metals [53, 55]: *1*) the mean of the data for sodium, potassium, rubidium and cesium, *2*) the data for lithium.

	Li	Na	K	Rb	Cs
$R_0 \cdot 10^{10}$, m³/C	−2.0 [113]	−2.3 [113]; −2.48 [52]	−4.2 [113]; −4.85 [52]	−5.9 [113]; −5.84 [52]	−7.8 [113]; −7.48 [52]
R_{liq}/R_0	—	0.98; 0.97	0.97	0.7; 0.97	0.96

Study [50] points out that the Hall coefficient of metals with approximately spherical Fermi surfaces depends slightly on temperature. The change is actually caused only by the anisotropy in the inelastic electron-phonon interaction which, in turn, leads to the development of an anisotropic angle structure and a fine energy structure in the nonequilibrium electronic distribution function. According to the calculations made in [50], the deviation of the Hall coefficient of sodium from the value typical for free electrons, amounts to only several percent.

§2. COPPER, SILVER, GOLD

COPPER

The shape of the Fermi surface and the band structure of copper is rather well established. The Fermi surface of copper has been investigated in a considerable number of experimental and theoretical studies; they are listed in [41]. It differs substantially from a spherical surface (which would follow from the free-electron model of a metal with one conduction electron per atom). The Fermi surface of copper touches the boundry of the Brillouin zone near the L point and, therefore, is multiply-connected (Figs. 13 and 14).

The curve of the density of electronic states $N(\varepsilon)$ was investigated extensively in [58–66]. The results of these studies show that the d-band of copper is located at a distance of ~1.5 eV from the Fermi level, and that it can sometimes influence the physical processes taking place in copper. This explains, for example, the variable valence of copper.

The elastic constants of copper ($c_{ij} \cdot 10^{-11}$ Pa) are listed below [61, 62]

T,K	4.2	80	200	300	400	500	600	700	800
c_{11}	1.745	1.742	1.716	1.700	1.655	1.615	1.575	1.535	1.495
c_{12}	—	—	—	1.225	1.205	1.185	1.165	1.145	1.125
c_{44}	0.819	0.814	0.787	0.758	0.731	0.704	0.677	0.650	0.623

At $T = 300$ K the listed values of c_{ij} differ from the data of other investigations by ±7% [61, 62].

At room temperature, polycrystalline copper has the following values for E, G and μ [36]
$E = 1.326 \cdot 10^{11}$ Pa; $G = 0.4925 \cdot 10^{11}$ Pa; $\mu = 0.35$

Figure 15 and Table 10 give $\rho(T)$, the electrical resistivity of copper, as a function of temperature. Electrical resistivity has long been carefully investigated for all temperatures. The uncertainty in the given values is ~1%. The compilation by Matula [63] provides the complete bibliography of the resistivity studies for copper (and also for silver and gold) published before 1977. Matula carefully discussed all of the data and recommended a dependence of $\rho(T)$, corrected for the thermal expansion. All kinetic property data refer to a metal containing less than 0.01% of impurities. The diffusivity of copper is also given in Table 10.

Figure 16 shows the temperature dependence of the thermal conductivity of copper. The uncertainty in the given values is 5% [48]. Figure 16 also shows the phonon component of thermal conductivity (λ_g), calculated in [23]. At 700 K it amounts to only 2–3% and decreases with the increase in temperature. Figure 16 also shows the results of the calculation of the electronic component from the standard W–F–L law. It is evident that above 500 K $\lambda_e^L \approx \lambda$ within the experi-

Figure 13 The Fermi surface of copper [41].

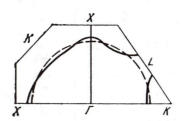

Figure 14 The distortions of the Fermi surface of copper compared against the Fermi surface of free electrons (broken line) [41].

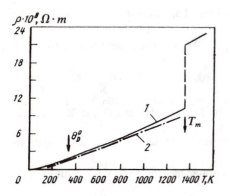

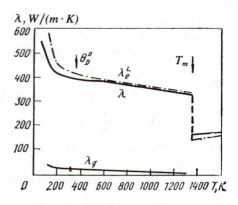

Figure 15 The electrical resistivity of copper (ρ) vs. temperature: *1*) [63]; *2*) calculations from the Bloch-Gruneisen theory [64–66].

Figure 16 The temperature dependence of the thermal conductivity of copper (λ) [48]: λ_g is the calculation of [23].

TABLE 10 The diffusivity (a) [39], electrical resistivity (ρ) [63] and hall coefficient (R) [68] for copper, silver and gold

T, K	$a \cdot 10^6$, m²/s			$\rho \cdot 10^8$, Ω·m			$R^* \cdot 10^{11}$, m³/C		
	Cu	Ag	Au	Cu	Ag	Au	Cu	Ag	Au
20	—	30 900	4830	—	—	—	—	—	—
40	—	1270	463	—	—	—	—	—	—
50	—	—	—	0.0518	0.144	0.221	—	—	—
80	—	268	173	—	—	—	—	—	—
100	210	227	155	0.348	0.418	0.650	—5.20	—8.9	—7.10
150	—	192	140	—	—	—	—	—	—
200	130	181	135	1.046	1.029	1.462	—5.25	—8.7	—7.17
250	—	176	131	—	—	—	—	—	—
273	—	175	130	—	—	—	—	—	—
300	117	174	128	1.725	1.629	2.271	—5.31	—9.0	—7.20
400	111	170	123	2.402	2.241	3.107	—5.37	—9.15	—7.25
500	—	—	—	—	—	—	—5.56	—9.3	—7.30
600	103	161	115	3.792	3.531	4.875	—5.63	—9.45	—7.33
700	—	—	—	—	—	—	—5.71	—9.60	—7.37
800	96.3	149	107	5.262	4.912	6.808	—5.82	—9.75	—7.40
900	—	—	—	—	—	—	—5.93	—9.9	—7.45
1000	90.3	137	99	6.858	6.396	8.986	—6.09	—10.12	—7.5
1100	—	—	—	—	—	—	—6.20	—10.2	—7.6
1200	80.6	124	91	8.626	8.089	11.49	—6.38	—10.3	—7.7
1300	—	—	—	—	—	—	—7.98	—12.0	—7.88
1400	42.7	—	—	21.01	18.69	31.08	—8.00	—	—11.80
1600	45.2	—	—	23.42	20.38	34.83	—	—	—
1800	47.4	—	—	—	—	—	—	—	—

*R_{sol}/R_{liq} amounts to 0.79, 0.85 and 0.67 for copper, silver and gold, respectively.

mental uncertainty. At lower temperatures $\lambda_e^L > \lambda$ because of inelastic scattering, which plays a larger role as the temperature decreases.

Figure 17 shows the temperature dependence of the absolute thermoelectric power of copper [67]. It is interesting to note that its value is positive and increases with temperature, contradicting the free-electron model. This behavior of $S(T)$ can be explained by the fine details of the Fermi surface.

Table 10 also lists the Hall coefficients for copper [67], silver and gold.

A model of nearly-free electrons, developed by Ziman [2], is frequently used for theoretical discussions of the kinetic properties of noble metals such as copper, silver and gold. According to this model, the electrical resistance of most of the monovalent metals is described by the well known Bloch-Gruneisen formula [2]. Using a matrix element according to the Bardeen formula [2] and neglecting the Umklapp processes, equation (7) describes, quite well, the temperature dependence of $\rho(T)$ for copper, silver and gold [64–66] (see Fig. 15).

Accurate numeric values of conductivity, however, cannot be obtained from the Bloch-Gruneisen theory [64–66].

Over the last ten years there have appeared theoretical investigations which use the pseudopotential method for calculating the band structure, the constants of the electron-phonon interaction, and the transport properties of noble metals in the solid and liquid states [64–66, 70, 71]. The authors of [66], for example, used the spherical Fermi surface model and proposed a new nonlocalized pseudopotential model for the noble metals. The theoretical estimations agree with the experimental results within 10–20%.

The temperature dependence of the absolute thermoelectric power of copper is shown in Fig. 17. The low-temperature anomaly is due to the effect of the electron-phonon entrainment. At $T > 273$ K the thermoelectric power is $S \cdot 10^6 = 1.722 + 0.00534 \, (T - 273)$ V/K [69]. The Hall coefficient of copper is also given in Table 10.

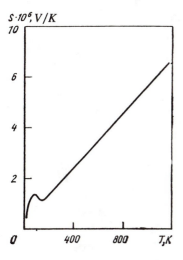

Figure 17 The absolute thermoelectric power (S) of copper vs. temperature [69, 67].

SILVER

At standard pressure silver has an fcc structure with the parameter $a = 0.40862$ nm at 298 K. The Fermi surface and band structure of silver are similar to those of copper (see Fig. 13). Open orbits, however, have smaller neck radii [41].

The curve of the density of electronic states was investigated in [70, 71]. The distance from the top of the d-band to the Fermi level amounts to 3 eV which is substantially greater than in copper or gold.

The elastic constants of silver are given in Table 11. These data differ by about 10% from the values of c_{ij} obtained by other investigators. The elastic characteristics of polycrystalline silver at room temperature are given in [73]: $E = 7.44 \cdot 10^{10}$ Pa; $G = 2.71 \cdot 10^{10}$ Pa; $\mu = 0.37$.

The temperature dependence of the electrical resistivity of silver is similar to that of copper (Fig. 18) [63]. The uncertainty in the given data are equal to $\approx 1\%$. The kinetic properties were measured on samples containing less than 0.01% impurities [63]. The temperature dependence of resistivity is described well by the Bloch-Gruneisen approximation (see Fig. 18). As in the case of copper, however, the numerical values of conductivity, calculated from the theory, differ from the experimental values by tens of percent [66].

The diffusivity of silver is given in Table 10 [3, 9]; the uncertainty of the listed values is about 5%.

The temperature dependence of the thermal conductivity of silver (Fig. 19) is similar to those of copper and gold (see Figs. 16 and 22). As for copper, the

Table 11   The elastic constants of silver ($c_{ij} \cdot 10^{-11}$, Pa)

T, K	c_{11}	c_{12}	c_{14}
According to data of [72]			
0	1.3149	0.9733	0.5109
50	1.3112	0.9726	0.5072
100	1.2980	0.9666	0.4982
150	1.2835	0.9591	0.4890
200	1.2691	0.9517	0.4797
250	1.2546	0.9444	0.4704
300	1.2399	0.9367	0.4612
According to data of [62]			
300	1.240	0.940	0.465
400	1.205	0.920	0.446
500	1.170	0.900	0.426
600	1.135	0.880	0.407
700	1.100	0.860	0.388
800	1.065	0.845	0.369

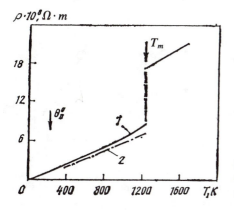

Figure 18 The electrical resistivity of silver (ρ) vs. temperature [63]: *1*) the recommended data; *2*) the calculation from the Bloch-Gruneisen theory.

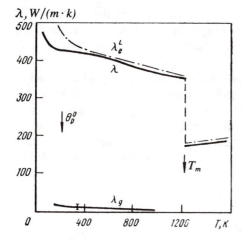

Figure 19 The thermal conductivity of silver (λ) vs. temperature [39, 48]; λ_g is the estimation from [23].

fraction of the lattice contribution does not exceed several percent, and it could not be noticed because of the 3–5% uncertainty in λ [48]. For silver, $\lambda_e^L > \lambda$ (see Fig. 19) below 400 K because the processes of inelastic scattering of electrons begin playing a larger role. This is a result of the difference in the relaxation times of the electrical resistivity and the thermal conductivity. The temperature dependence of the absolute thermoelectric power of silver (*S*, μV/K) is similar to that of copper (Fig. 20). The Hall coefficient of silver is given in Table 10.

GOLD

At standard pressure, gold has an fcc structure with the parameter $a = 0.40785$ nm. Its electronic properties are similar to those of silver and copper, although the deviation of the Fermi surface from a sphere is somewhat greater than in either of them. .

The elastic constants of gold are listed in Table 12 [62, 72]. At room tem-

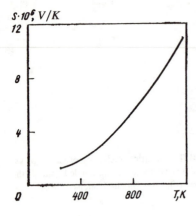

$S \cdot 10^6, V/K$

Figure 20 The absolute thermoelectric power of silver (S) vs. temperature [69].

perature, other data from numerous investigations differ from the data given in Table 12 by no more than $\pm 5\%$. At room temperature, the elastic characteristics of polycrystalline gold are equal to: $E = 7.72 \cdot 10^{10}$ Pa; $G = 2.71 \cdot 10^{10}$ Pa; $\mu = 0.42$ [73].

The temperature dependence of the electrical resistivity of gold is shown in Fig. 21. The kinetic properties were determined from the samples containing usually less than 0.005% impurities. The uncertainties in the given values are about 1–2%. Gold is characterized by higher values of electrical resistivity than silver or copper and a temperature coefficient above 400 K. The experimental values of $\rho(T)$ deviate quite visibly from the values, calculated on the basis of the Bloch-Gruneisen approximation.

Table 12 The elastic constants of gold ($c_{ij} \cdot 10^{-11}$, Pa)

T, K	c_{11}	c_{12}	c_{44}
According to data of [72]			
0	2.0163	1.6967	1.4544
50	2.0107	1.6937	0.4510
100	1.9940	1.6822	0.4446
150	1.9763	1.6693	0.4384
200	1.9587	1.6567	0.4320
250	1.9410	1.6440	0.4257
300	1.9234	1.6314	0.4195
According to data of [62]			
300	1.925	1.630	0.424
400	1.890	1.605	0.410
500	1.850	1.580	0.597
600	1.820	1.555	0.383
700	1.785	1.530	0.369
800	1.755	1.505	0.355

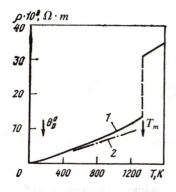

Figure 21 The electrical resistivity of gold (ρ) vs. temperature: *1*) the average values [63]; *2*) the values calculated using the Bloch-Gruneisen theory.

The diffusivity of gold is listed in Table 10; the uncertainty in the values is about ~5%. Similar to copper and silver, at temperatures above 100 K, gold has $\partial a/\partial T < 0$ in the solid state and $\partial a/\partial T > 0$ in the liquid state. The temperature dependence of the thermal conductivity of gold is similar to that of copper and silver, although the values of λ are somewhat smaller.

The uncertainty in the listed data (Fig. 22 and Table 13) [48] is 2% at moderate temperatures and about 5% at the extremes of the temperature interval. Above

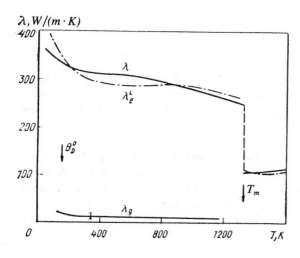

Figure 22 The thermal conductivity of gold (λ) vs. temperature [48]: $\lambda_e^L = L_0 T/\rho$; λ_g is the estimation according to [23].

Table 13 The thermal conductivity (λ) of copper, silver, gold, W/(m · K) [48].

T, K	Cu	Ag	Au	T, K	Cu	Ag	Au
100	483	475	360	800	371	397	284
200	413	441	326	1000	357	383	271
300	398	433	315	1200	342	—	—
400	392	426	309	1400	167	—	—
600	383	411	296	1600	174	—	—

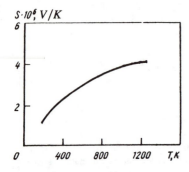

$S \cdot 10^6$, V/K

Figure 23 The absolute thermoelectric power (S) of gold vs. temperature [69].

300 K the electronic component of the thermal conductivity is almost equal to the total thermal conductivity, and below 150 K $\lambda_e^L > \lambda$.

The temperature dependence of the absolute thermoelectric power of gold (S, μV/K) is similar to that of copper and silver (Fig. 23). According to the data of [69], $S = 1.72 + 4.6 \cdot 10^{-3} (T - 273) - 2.25 \cdot 10^{-6} (T - 273)^2$. The Hall coefficient of gold is given in Table 10.

§3 BERYLLIUM, MAGNESIUM, CALCIUM, STRONTIUM, BARIUM, RADIUM

BERYLLIUM

At standard pressure beryllium has two crystalline modifications: an hcp phase, α-Be, with the parameters $a = 0.2286$ nm and $c = 0.3584$ nm at 293 K, and a bcc phase, β-Be, with the parameter $a = 0.2551$ nm at 1528 K. The temperature of the α-β transformation is 1533 K according to the data of [74], and 1528 K according to [39].

The electronic structure of beryllium has been well studied [41]. Its Fermi surface consists of two principal sheets: the region of holes in the second Brillouin zone similar to a 'coronet', and two identical pockets of electrons in the third zone similar to a 'cigar' with a triangular cross section.

The elastic constants ($c_{ij} \cdot 10^{-11}$, Pa) of 99.1% pure beryllium are given below [75].

T, K	298	323	373	423	473	523	573
c_{11}	2.888	2.859	2.802	2.745	2.689	2.635	2.583
c_{33}	3.542	3.515	3.457	3.402	3.353	3.307	3.251
c_{44}	1.549	1.540	1.522	1.505	1.487	1.471	1.454
c_{12}	0.201	0.196	0.185	0.174	0.165	0.155	0.146
c_{13}	0.047	0.037	0.016	−0.003	−0.022	−0.040	−0.058

The temperature dependence of the electrical resistivity of beryllium is given in Fig. 24 in accordance with the assessment of more than 80 experimental works in [76].

The kinetic properties of beryllium are listed below [39, 48, 76, 77]

T,K	100	200	300	400	500	600
$\rho \cdot 10^8, \Omega \cdot m$	0.133	1.29	3.76	6.76	9.94	13.2
$a \cdot 10^6, m^2/s$	—	146	59.0	39.8	31.5	26.9
$\lambda, W/(m \cdot K)$	—	330	215	170	140	120

T,K	700	800	900	1000	1200	1400
$\rho \cdot 10^8, \Omega \cdot m$	16.5	20.0	23.7	27.5	35.7	44.8
$a \cdot 10^6, m^2/s$	23.5	20.9	18.7	16.8	13.8	11.5
$\lambda, W/(m \cdot K)$	110	100	92	90	85	—

The discrepancies between the above values and the literature data for poly-crystalline beryllium, given in [77], reach 200%. This might be the result of in-sufficient purity of samples and of the influence of anisotropic factors, since the anisotropy of resistivity of beryllium reaches 40%. The data given above apply to a metal of 99.9% purity, with a relative residual resistance of $r = 107$ for a polycrystalline sample, 400 for $\rho_\perp c$, and 900 for $\rho_\| c$.

The Debye temperature of beryllium is rather high (at $T = 0$ $\theta_D = 1160$ K), so for beryllium the region below 1000 K is the low-temperature region. Therefore $\rho\,(T)$ can be described rather well by the ordinary Bloch-Gruneisen expression [77]. Some increase in its curvature is probably due to the thermal expansion and consequently to the resulting change in the phonon spectrum (the decrease in the Debye temperature according to the Bloch-Gruneisen theory, at 298 K $\theta_D = 1031$ K).

The diffusivity of beryllium decreases rapidly as temperature increases. The values given for the polycrystalline metal have a 25% uncertainty (39).

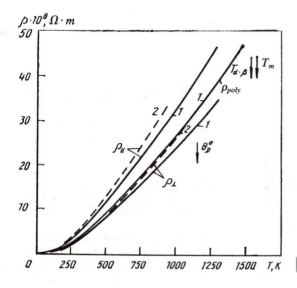

Figure 24 The electrical resistivity of beryllium (ρ) vs. temperature: *1*) [77]; 2) [76].

Beryllium is a good conductor. At 150 K its thermal conductivity is comparable to the thermal conductivity of copper (Fig. 25) [48]. Overall, the dependence of $\lambda(T)$ has a negative temperature coefficient. Within the framework of the W–F–L law, $\lambda_e \approx \lambda$, and according to the calculations of [23], the fraction of λ_g does not exceed 9% if the phonon-electron scattering is taken into account.

Above 200 K the absolute thermoelectric power of beryllium is positive and has a positive temperature coefficient similar to the case of other alkaline-earth metals. The curve in the Fig. 26 corresponds to the average dependence of $S(T)$ for the metals of this group [55].

MAGNESIUM

At standard pressure magnesium has an hcp structure with the parameters $a = 0.32094$ nm and $c = 0.52103$ nm at 298 K [74]. The electronic structure of the Fermi surface of magnesium has been well established [41]. It should be pointed out that the Fermi surface of magnesium is fairly similar to the free-electron model for the divalent metals with an hcp structure. In general it is similar to the Fermi surface of beryllium as well as zinc and cadmium. The elastic constants ($c_{ij} \cdot 10^{-10}$, Pa) of magnesium are given below [78].

$T, K \cdots$	0	80	160	240	300
$c_{11} \cdots$	6.348	6.300	6.189	6.049	5.940
$c_{12} \cdots$	2.594	2.591	2.585	2.573	2.561
$c_{13} \cdots$	2.170	2.168	2.163	2.155	2.144
$c_{33} \cdots$	6.645	6.595	6.455	6.281	6.16
$c_{44} \cdots$	1.842	1.820	1.768	1.697	1.64
$c_{66} \cdots$	1.875	1.858	1.801	1.738	1.690

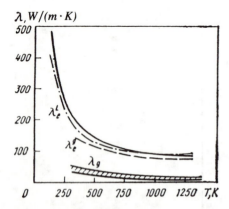

Figure 25 The thermal conductivity of beryllium (λ) vs. temperature [48]: λ_g is the calculation of [23]; $\lambda_e^g = \lambda - \lambda_g$; $\lambda_e^L = L_0 T/\rho$, where ρ comes from the estimations of [77].

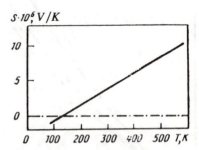

Figure 26 The generalized temperature dependence of the thermoelectric power (S) for ordinary metals of the Second group [55].

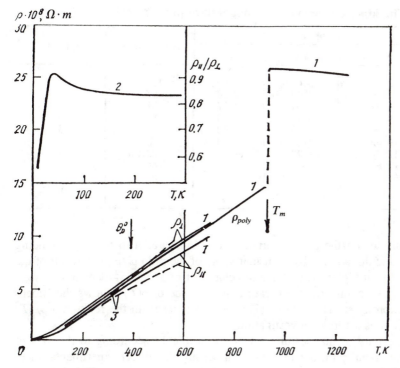

Figure 27 The temperature dependence of the electrical resistivity of single crystal magnesium in the direction of the hexagonal axis ($\rho_\parallel$), perpendicular to it ($\rho_\perp$), and of polycrystalline magnesium (ρ_{poly}): *1*) [77]; *2*) [79]; *3*) [80].

The data given above are for the samples of 99.9% purity. The uncertainty in the values of constants c_{11}, c_{33} and c_{44} does not exceed 0.70%, and for constants c_{12} and c_{13} it amounts to 1.8%.

Magnesium is characterized by a strong temperature dependence of the shear constants c_{14} and c_{66} against a slight temperature dependences of the compression constants c_{11} and c_{33}. Although the lattice parameters are anisotropic, which is the usual case for metals with an hcp structure ($c/a = 1.623$), the moduli ratio, which describes the elastic anisotropy, is close to one (c_{66}/c_{44} and c_{33}/c_{11} are equal to 1.018 and 1.047 at 0 K, and to 1.030 and 1.037 at 300 K, respectively).

Figure 27 and Table 14 give the data on the temperature dependence of the electrical resistivity of magnesium. The data in [77] refer to a metal with $r = 700$ for a polycrystal and $r \approx 470$ for a single crystal; their uncertainties are 3% in the interval 100–600 K, 5% from 600 to the α-β transformation point, and ~10% at higher temperatures. Figure 27 also shows the data in [79], which discussed the anisotropy of resistivity. After reaching a maximum near 40 K, the ratio $\rho_\parallel$/$\rho_\perp$ becomes practically constant for temperatures above 150 K. Studies [8, 79] indicate that at room temperatures and higher, the anisotropy of resistivity is determined by the ratio $S_\perp/S_\parallel$, where $S_\perp$ and $S_\parallel$ are the areas of the projection of

Table 14 The kinetic properties of magnesium [39, 48, 77]

T, K	$\rho \cdot 10^8$, $\Omega \cdot m$			$a \cdot 10^6$, m^2/s	W/(m·K)
	Polycrystal	∥ c	⊥ c		
100	0.908	0.827	0.983	148	—
200	2.75	2.42	2.90	97.1	—
300	4.51	3.94	4.67	87.4	156
400	6.19	5.42	6.39	82.8	—
500	7.86	6.90	8.09	79.2	151
600	9.52	8.35	9.76	75.6	—
700	11.2	9.78	11.4	72.2	—
800	12.8	—	—	68.9	—
900	14.4	—	—	65.6	—
1000	—	—	—	—	84

the Fermi surface on the planes perpendicular and parallel to the corresponding crystallographic directions. The constant value of the ratio $\rho_{\parallel}/\rho_{\perp}$ is a result of the assumption that at high temperatures the character of $S_F^{\parallel}$ and $S_F^{\perp}$ does not change.

Generally speaking, the temperature dependence of $\rho(T)$ follows the Bloch-Gruneisen expression. The data of [77], however, result in the relation $\partial^2\rho/\partial T^2 < 0$; this requires a further investigation.

The diffusivity of magnesium is given in Table 14 according to the data of [39]. The data require further investigation because the direct experimental measurements of the values of a are absent, and the anisotropy in diffusivity is unknown. The uncertainties in these data are estimated as being 13% [39].

The information on the thermal conductivity of polycrystalline magnesium (Fig. 28 and Table 14) is preliminary. The uncertainty is of the same order of magnitude as the anisotropy of the thermal conductivity ($\approx$15%). This can be estimated from the data for ρ. At temperatures above 100 K, the thermal conductivity of magnesium is mostly of the electronic nature $\lambda \approx \lambda_e$, and the fraction due to the lattice component does not exceed 3% at $T = 2\theta_D^{\circ}$ [23].

The temperature dependence of the absolute thermoelectric power of poly-

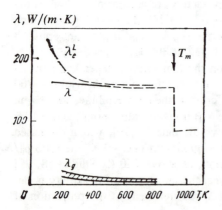

Figure 28 The thermal conductivity of magnesium (λ) vs. temperature [48]; $\lambda_e^L = L_0T/\rho$; λ_g [23].

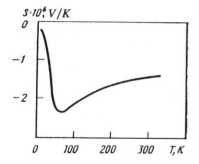

$S \cdot 10^6, V/K$

Figure 29 The thermoelectric power of magnesium vs. temperature [81].

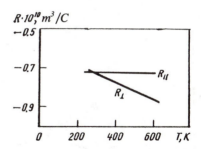

$R \cdot 10^{10}, m^3/C$

Figure 30 The Hall coefficient (R) of magnesium in the direction of the hexagonal axis ($R_{\parallel}$) and perpendicular to it ($R_{\perp}$) as a function of temperature [80].

crystalline magnesium has a minimum near 70 K, and at higher temperatures it has a positive slope (Fig. 29).

Fig. 30 shows the Hall coefficient of magnesium at moderate temperatures [80]. At 300 K, $R_{\perp} \approx R_{\parallel} = -(0.73 \pm 0.1) \cdot 10^{-10}$ m³/C.

CALCIUM

At standard pressure calcium has two crystalline modifications: an hcp phase, α-Ca, with the parameter $a = 0.55884$ nm at 299 K, and a bcc phase, β-Ca, with $a = 0.448$ nm at 740 K. The transformation temperature is 737 K according to [39], or 740 K according to [74].

The existence of small pockets of electrons in the second zone leads to a strong dependence of the resistance on pressure.

The information about the elastic properties of polycrystalline calcium for the temperature interval 93–873 K, and also on its Young's modulus at various temperatures, is given below [78, 82, 85].

	Ca	Ba	Sr
$E \cdot 10^{-10}$, Pa $\cdots\cdot$	2.0	1.6*	1.29
$G \cdot 10^{-10}$, Pa $\cdot\cdot\cdot$	0.75	0.62*	0.5
$K \cdot 10^{-10}$, Pa $\cdots$	1.76	1.23	0.98
$\mu \cdots\cdots\cdots\cdots$	0.31	0.28*	0.28

*The calculated values.

T, K $\cdots\cdots\cdots$	93	293	473	673	873
$E \cdot 10^{-11}$, Pa $\cdots$	20.6	19.6	17.7	15.7	12.3

Figure 31 shows the resistivity of calcium as a function of temperature. The data, combined in [77], have an uncertainty, below 300 K, of 5%, and of ~20%

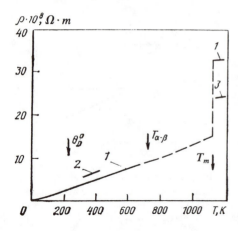

Figure 31 The electrical resistivity of calcium (ρ) vs. temperature: *1*) [77]; *2*) [83]; *3*) [84] (above 600 K the values shown were taken from [77]).

at higher temperatures. They are given for the metal of 99.96% purity with $r = 70$. Above 300 K, these data require verification. On the whole, the temperature dependence of $\rho(T)$ is close to that which can be expected from the Bloch-Gruneisen formula.

The temperature dependent kinetic properties of calcium are given below [39, 40, 77].

T,K $\cdots\cdots\cdots$	100	200	300	400	500	600
$\rho\cdot10^{8},\Omega\cdot$m $\cdots$	0.913	2.19	3.45	4.73*	6.02*	7.35*
$a\cdot10^{8}$,см²/s $\cdots$	—	2.30	1.99	1.78*	1.63*	1.52*
T,K $\cdots\cdots\cdots$	700	800	900	1000	1100	
$\rho\cdot10^{8},\Omega\cdot$m $\cdots$	8.70*	10.0*	11.4*	12.8*	14.3*	
$a\cdot10^{8}$,см²/s $\cdots$	1.43*	1.25*	0.942*	0.810*	—	

*The data need revision.

The diffusivity data for calcium have an uncertainty of about 30%. Its thermal conductivity is shown in Fig. 32, and has a considerable uncertainty. The thermal conductivity of calcium is largely electronic in nature. The differences between the data of [48] and [83] are probably due to the influence of impurities.

The thermoelectric power of calcium is positive, and above room temperature, it increases (Fig. 33). In the liquid state, however, $\partial S/\partial T < 0$ [84].

STRONTIUM

At standard pressure below 488 K, strontium has an fcc structure with the parameter $a = 0.60849$ nm at 298 K [74]. A number of reference books [39, 77, 82] have stated that between 488 and 815–830 K, it transforms into an hcp modification. However, study [74] showed that the fcc structure is stable up to 830 K, when it transforms into a bcc structure with the parameter $a = 0.485$ nm at 887

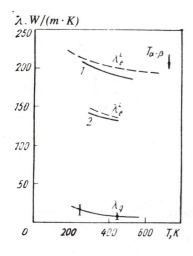

Figure 32 The thermal conductivity of calcium (λ) vs. temperature: $1)$ = [48]; for $\lambda_e^L = L_0 T/\rho$; ρ from [77]; $2)$ [83] and λ and $\lambda_e^L = LT/\rho$ the estimation of the values of ρ was taken from the data of [83].

K. The electronic structure of fcc strontium is similar to that of calcium. The pockets of holes and electrons, however, are substantially smaller in size [41].

The information about the elastic constants of strontium [85] is limited and is only of a preliminary nature. The temperature dependence of the electrical resistivity of strontium is shown in Fig. 34. The data of [76, 77] do not show any changes in the electrical resistivity near the assumed fcc-hcp transition point, therefore, the probability that an hcp phase exists is small.

The kinetic properties of strontium are given below [39, 77]

$T, K \cdots$	100	200	300	400	500	600	700	800	900	1000
$\rho \cdot 10^8, \Omega \cdot m \cdots$	4.58	9.04	13.5	17.8	22.2	26.7	31.2	35.6	54*	62*
$a \cdot 10^6, M^2/s \cdots$	—	—	45.2	39.0	32.6	31.0	30.0	29.6	23.7	23.3

*The data require further investigation.

The above data describe a metal of 99.95% purity with $r \approx 16$. The uncertainty amounts to 5% in the interval from 50 to 815 K, 10% up to the melting

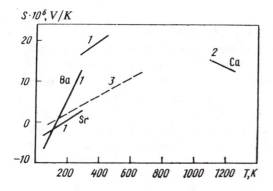

Figure 33 The thermoelectric power of Ba, Ca, Sr vs. temperature [55, 83]: $1)$ = [83]; $2)$ [84]; $3)$ the combined dependence for the metals of the second group according to the data of [55].

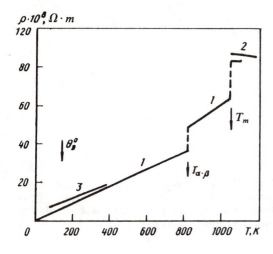

Figure 34 The electrical resistivity of strontium (ρ) vs. temperature: *1*) [77]; *2*) [84]; *3*) [86].

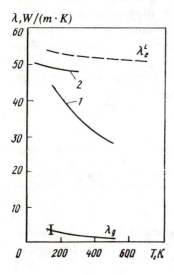

Figure 35 The thermal conductivity of strontium (λ) vs. temperature: *1*) [48]; *2*) [83], $\lambda_e^L = L_0 T/\rho$ (ρ from [77]); λ is the estimations of [23].

point, and 20% for the liquid phase [17]. The dependence of $\rho(T)$ is similar to that expected from the Bloch-Gruneisen theory [86]. The information on diffusivity of strontium [39] is only preliminary because direct experimental investigations are absent. The uncertainty of these data is estimated as 25%. The thermal conductivity of strontium (Fig. 35) also requires further investigation. The data of [83] apply to a metal with $r = 15.5$. For these data $\lambda \approx \lambda_e$, although it is possible that the Lorenz function was underestimated. The thermoelectric power of strontium has a positive temperature coefficient and an inversion point near 600 K [55, 83] (see Fig. 33).

BARIUM

At standard pressure barium has a bcc structure with the parameter $a = 0.5013$ nm at 298 K [74]. The structure of the Fermi surface of barium has a certain resemblance to the surfaces of the fcc calcium and strontium.

The following are the known elastic parameters of barium, given at $T = 293$ K [87]: $v_e = 2235$ m/s, $v_t = 1325$ m/s, $E = 1.585 \cdot 10^{-10}$ Pa, $G = 0.646 \cdot 10^{10}$ Pa, $K = 0.976 \cdot 10^{10}$ Pa, $\mu = 0.229$.

Figure 36 shows the temperature dependence of the electrical resistivity of barium. The data of [77] are given for a metal of 99.5% purity and with the relative residual resistance of $r = 40$. The uncertainty in the above values is 5% below 700 K, and 10% at higher temperatures. It should be noted that the electrical resistivity of barium reaches the highest values of any metal, 300 $\mu\Omega \cdot$ cm, and its temperature dependence [77] is characterized by a steep positive slope in the solid state and a negative slope in the liquid state. At $T < T_m$ the data of [88] differ substantially from those of [77]. The electrical conductivity of barium is given below [39, 77].

T, K	100	200	300	400	500	600	700	800	900	1000
$\rho \cdot 10^8, \Omega \cdot$ m	8.85	20.2	34.3	51.4	72.4	98.2	130	168*	216*	275*

*The data require further investigation.

The information on diffusivity of barium is rather scarce. It is known that $a = 31.4 \cdot 10^{-6}$ m^2/s, at 200 K, and $27.4 \cdot 10^{-6}$ m^2/s at 300 K. Even for this narrow temperature interval, the data are of a preliminary nature. Their uncertainty is ~25% because direct experimental data are absent.

The information on the thermal conductivity of barium is also limited. It is known that $\lambda = 29.0$ W/(m$\cdot$K) at 100 K, 24.5 W/(m$\cdot$K) at 200 K, and 20.5 W/(m$\cdot$K) at 300 K. According to the data of [83], at room temperature $\lambda_e \approx \lambda$. At lower temperatures, however, it is possible that $\lambda_e < \lambda$.

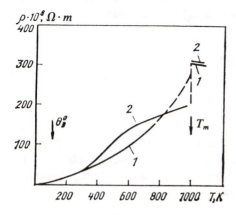

Figure 36 The electrical resistivity of barium (ρ) vs. temperature: 1) [77]; 2) [88].

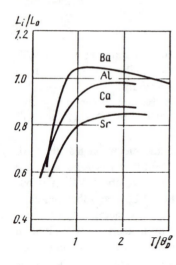

Figure 37 The reduced Lorenz function L/L_0 vs. the reduced temperature for barium, calcium, strontium and aluminum [83].

Figure 37 shows the graph of the Lorenz function vs. the reduced temperature T/θ_D° for calcium, strontium, barium, and aluminum.

A small excess in λ over λ_e can be expected for barium due to the lattice contribution; for all other metals $\lambda_e > \lambda$. In addition, L is substantially smaller than L_0 because of the inelastic nature of the electron scattering at temperatures below $0.5\theta_D$. This phenomenon was also observed for the alkali metals.

The thermoelectric power of barium has a positive temperature coefficient and an inversion point near 140 K (see Fig. 33).

RADIUM

Radium does not have a stable isotope and its physical properties are very poorly known. Study [77] indicates that it has a melting point at 973 K. The electrical resistivity was estimated at 88 $\mu\Omega \cdot$ cm at 300 K, 145 $\mu\Omega \cdot$ cm at 400 K, and 212 $\mu\Omega \cdot$ cm at 500 K. The uncertainty in the above values is estimated as 80% [77].

§4 ZINC, CADMIUM, MERCURY

ZINC

At standard pressure zinc has an hcp structure with the parameters $a = 0.26649$ nm and $c = 0.49468$ nm at 298 K. The ratio $c/a = 1.856$ considerably exceeds the value of 1.633, which is the ideal ratio for the hcp lattice. The hcp structure of zinc remains stable up to the melting point [74].

The structure of the Fermi surface of zinc is well established [41]. It is similar to the free electron surface of a divalent hcp metal. The deviation, however, of the lattice parameter ratio from the ideal value and spin-orbital interaction lead to some pronounced new features [41]. The Fermi surface of zinc is characterized by substantial open surfaces of holes in the first and second zones and closed sheets of electrons in the third and fourth zones. It should be pointed out that the Fermi surface of cadmium is very similar to that of zinc. The elastic constants of zinc ($c_{ij} \cdot 10^{-11}$, Pa) are listed below [89].

T, K	4.2	77	150	200	300	400	500	600	670
c_{11}	1.7909	1.7677	1.7336	1.7047	1.6368	1.590	1.4648	1.3843	1.3395
c_{12}	0.375	0.368	0.366	0.367	0.364	0.358	0.351	0.365	0.408
c_{13}	0.554	0.552	0.548	0.545	0.530	0.515	0.508	0.504	0.502
c_{33}	0.6880	0.6766	0.6620	0.6523	0.6347	0.6168	0.5982	0.5793	0.5661
c_{44}	0.4595	0.4479	0.4296	0.4158	0.3879	0.3573	0.3261	0.2933	0.2667

The temperature dependence of these constants has a specific feature different from the other hcp metals: the shear elastic constants change more than the longitudinal ones. In addition, zinc, as cadmium, has a high anisotropy of the elastic properties, which can be seen from the data given below. The ratios of constants c_{44}/c_{66}, c_{11}/c_{33} and $K_{\perp}/K_{\parallel}$ are compared below [89] for the hcp crystals of beryllium, magnesium, cadmium and zinc at $T = 0$ K

	Be	Mg	Cd	Zn
c/a^*	1.585	1.623	1.855	1.886
c_{44}/c_{66}	1.222	0.982	0.669	0.542
c_{11}/c_{33}	0.875	0.955	2.591	2.278
$K_{\perp}/K_{\parallel}$	1.09	0.96	0.13	0.18

*At $T = 300$ K.

Unlike the other hcp metals, zinc, above 100 K, has a surprisingly small anisotropy of the electrical resistivity (Fig. 38) [79]. The ratio $\rho_{\parallel}/\rho_{\perp}$ has a maximum at about 50 K. The Debye temperature of zinc is close to room temperature. At higher temperatures, the anisotropy of zinc is mainly determined by the anisotropy of its Fermi surface. Small values of the anisotropy of the electrical resistivity can be explained by the fact that the projections of the Fermi surface areas in the corresponding directions are approximately equal.

In general, the temperature dependence of the electrical resistivity of zinc is almost linear: $\rho = AT$, despite the fact that the temperature coefficient increases slightly with temperature.

The diffusivity of solid zinc decreases with the increase in temperature [39, 48, 79].

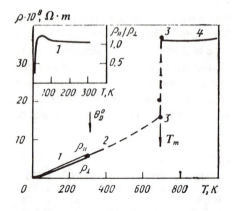

$\rho \cdot 10^8, \Omega \cdot m$

Figure 38 The electrical resistivity of zinc (ρ) vs. temperature: 1) [79]; 2) [104]; 3) [113]; 4) [105].

T, K	100	200	300	400	500	600	700	750
$\rho \cdot 10^8, \Omega \cdot m$ . . .	0.25	4.0	6.0	8.0	10.5*	13.0*	16.0*	37.5

T, K	100	200	300	400	500	700
$a \cdot 10^6, M^2/s$ . . .	55.0	44.8	41.6	38.9	36.5	15.8

T, K	200	300	400	500	600	700	800	900	1000	1100
$\lambda, W/(m \cdot K)$ . . .	122	121	116	110	105*	100*	55*	61*	67*	73*

*Preliminary data.

The data listed above describe polycrystalline zinc of 99.999% purity. The uncertainty amounts to about 8% for the solid state and 15% for the liquid state [39]. The thermal conductivity of zinc also has a negative temperature coefficient in the solid state and positive one in the liquid state. It is electronic in nature. The electronic component is almost equal to the total value, deviating by 10–15% in accordance with the standard W-F-L law. The uncertainty in the values given in Fig. 39 is estimated as 3% [48] at room temperature, and 10–15% at higher and lower temperatures. The information on the anisotropy of the thermal conductivity is absent; the available data are only preliminary [48].

The thermoelectric power of polycrystalline zinc increases during the temperature interval 300–600 K from 1.2 to 9 $\mu V/K$ [109]. At 293 K the Hall coefficient of zinc is equal to $1.4 \cdot 10^{-10}$ m^3/C for the direction parallel to the hexagonal axis and $-2.7 \cdot 10^{-10}$ m^3/C for the perpendicular direction [80]. This coefficient changes almost linearly with temperature, and at 550 K it is equal to 1.10^{-10} and $0.2 \cdot 10^{-10}$ for the two directions, respectively.

CADMIUM

Cadmium, as zinc, has an hcp structure with a high ratio of lattice parameters (at 294 K $a = 0.29788$ nm, $c = 0.56176$ nm, $c/a = 1.885$). The Fermi surfaces of cadmium and zinc are similar (Fig. 40).

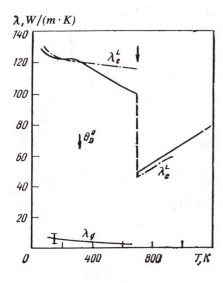

Figure 39 The thermal conductivity of zinc (λ) vs. temperature [48]: λ_s are the results calculated using equations (34) and (35).

The elastic constants of cadmium are listed in Table 15 [90, 91]. Cadmium has the same special features as zinc: a more rapid increase in the shear constants with temperature and a high elastic anisotropy.

For cadmium the anisotropy in the electrical resistivity is considerably higher than in zinc, despite the fact that the shape of the tensor components of resistivity vs. temperature do not change. The increase in the slope of resistivity with tem-

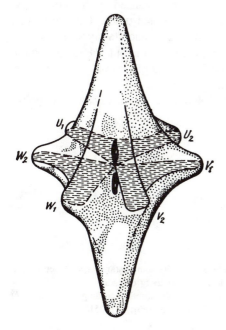

Figure 40 The Fermi surface of holes in the first zone of cadmium [41].

Table 15 The elastic constants of cadmium ($c_{ij} \cdot 10^{-10}$, Pa)

T, K	c_{11}	c_{12}	c_{13}	c_{33}	c_{44}	c_{66}
	According to data of [90]					
0	13.08	4.048	4.145	5.737	2.449	4.516
40	13.02	4.032	4.126	5.690	2.424	4.494
80	12.89	4.040	4.129	5.600	2.370	4.425
120	12.69	4.030	4.120	5.509	2.315	4.330
160	12.47	4.016	4.120	5.420	2.262	4.227
200	12.24	4.014	4.113	5.336	2.203	4.113
240	11.97	3.994	4.100	5.253	2.140	3.988
300	11.52	3.972	4.053	5.122	2.025	3.774
	According to data of [91]					
300	11.450	3.950	3.990	5.085	1.985	—
340	11.40	3.930	4.000	5.005	1.872	—
400	10.610	3.930	3.980	4.870	1.703	—
440	10.220	3.950	3.960	4.770	1.590	—
500	9.540	3.970	3.950	4.615	1.415	—
540	9.020	3.980	3.960	4.510	1.293	—
575	—	3.940	3.960	4.410	1.187	—

perature has been reliably established up to the melting point (Fig. 41 and Table 16).

A substantial decrease in the temperature coefficient in the liquid state is typical for these metals. Studies [79, 92] showed that near and above the Debye temperature, the anisotropy in the electrical resistivity is caused by the anisotropy of the Fermi surface (the time of relaxation becomes isotropic). Therefore, the anisotropy does not really depend on temperature. The uncertainty in the given values of ρ amounts to about 3% in the solid state and to about 5% in the liquid state. The data for $\rho_\perp$ and $\rho_\parallel$ refer to a 99.999% pure single crystal. The data for ρ_{poly} refer to a polycrystalline sample of similar purity.

The temperature dependence of the diffusivity of single crystal and poly-

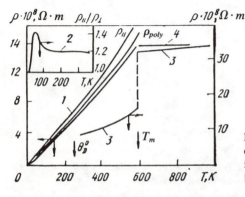

Figure 41 The temperature dependence of the electrical resistivity of cadmium (ρ) and its anisotropy (in the inset) according to the data: *1*) [79]; *2*) [92]; *3*) [96]; *4*) [105].

Table 16 The kinetic properties of cadmium [39, 48, 79, 92, 93, 113]

T, K	$\rho \cdot 10^8$, $\Omega \cdot m$			$a \cdot 10^8$, m^2/s			λ, $W/(m \cdot K)$		
	$\rho_\parallel$	$\rho_\perp$	ρ_{poly}	$a_\perp$	$a_\parallel$	α_{poly}	$\lambda_\perp$	$\lambda_\parallel$	λ_{poly}
100	2.6	2.2	2.4	—	—	—	110	88.0	103
200	5.3	4.3	4.7	54.5	44.0	50.5	106	85.0	99.0
300	8.0	6.9	7.2	51.5	41.5	48.0	102	82.0	97.0
400	11.3	9.5	10.0	49.0	39.0	46.0	100	81.0	94.0
500	14.4	12.1	13.0	46.0	37.0	43.5	99.0	79.0	91.5
600	—	15.25	—	44.0	35.0	42.0	—	—	88.5
700	—	—	33.0	—	—	17.0	—	—	50.0
800	—	—	33.5	—	—	17.0	—	—	57.0
900	—	—	34.0	—	—	16.0	—	—	—

crystalline cadmium, shown in Fig. 42, is typical for the nontransition hcp metals. In the solid state above 100 K, the diffusivity decreases while its anisotropy does not change; in the liquid state it depends slightly on temperature (or increases). The data of [79, 96] apply to high (99.999%) purity cadmium; their uncertainty is about 8% for the solid state and 10–15% for the liquid state.

The temperature dependence of the thermal conductivity along the principal axes of single crystals and polycrystalline cadmium is typical. The same can be said about cadmium in the liquid state. The thermal conductivity is mainly due to the electronic component; the lattice contribution was estimated as 2–4% of the total value of λ at $T = 2\theta_D$ [23]. Therefore, the electronic component coincides with the overall thermal conductivity within the uncertainty limit of about 5% at room temperatures and 15% for the liquid state (Fig. 43).

In the 300–600 K temperature interval, the thermoelectric power of poly-crystalline cadmium increases approximately linearly from 2 to 10 μV/K [109]. At 293 K, for the corresponding directions, the Hall coefficient of cadmium is equal to: $R_\parallel = (1.48 \pm 0.05)10^{-10}$ m^3/C; $R_\perp = (0.37 \pm 0.02)10^{-10}$ m^3/C. It changes approximately linearly with temperature, reaching the values of $R_\parallel = 0.98 \cdot 10^{-10}$ m^3/C and $R_\perp = 0.20 \cdot 10^{-10}$ m^3/C at 550 K [80].

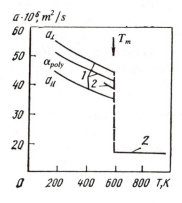

Figure 42 The diffusivity of cadmium (*a*) vs. temperature according to the data of *1*) [39]; 2) [51].

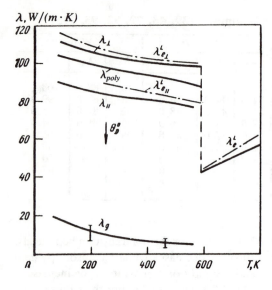

$\lambda, W/(m \cdot K)$

MERCURY

Mercury has the lowest melting point among metals (T_m = 234.288 K) [39]. Below the melting point, at standard pressure, mercury has a rhombohedral structure with the parameters a = 0.3005 nm, and α = 70.53° at 227 K. Study [74] pointed out that at 79 K a bct phase can appear with the parameters a = 0.3995 nm, c = 0.2825 nm at 77 K.

The Fermi surface of rhombohedral mercury is similar to that of free electrons; however, there are distortions because it intersects the faces of the Brillouin zone.

At 83 K, mercury has the following values for the elastic constants, in 10^{10} Pa [95]: c_{11} = 3.6; c_{12} = 2.89; c_{13} = 3.03; c_{14} = 0.47; c_{33} = 5.05; and, c_{44} = 1.29. The volume compressibility of mercury is equal to $3.7 \cdot 10^{-11}$ Pa at 293 K.

The kinetic properties of mercury (of 99.999% purity) are given in Figs. 44–46 and in Table 17 [39, 48, 96]. The anisotropy reaches 30%, and the conductivity is higher along the trigonal axis. During melting a significant jump in the electrical resistivity is typical for mercury (for a polycrystalline sample ρ_{liq}/ρ_{sol} = 4.2). The thermal conductivity of mercury is of an electronic nature. Generally speaking, the kinetic properties of mercury require further investigation, because the available values have a relatively high uncertainty (10–15%).

In the solid state, near the melting point, the thermoelectric power of mercury is S = −2 μV/K, and in the liquid state S = −3.5 μV/K [113]. In the solid state, near the melting point, the Hall coefficient of mercury is R = $-7.6 \cdot 10^{-11}$ m^3/C, and, during melting, it changes slightly.

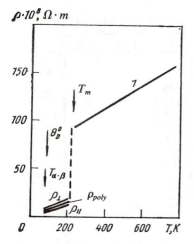

Figure 44 The electrical resistivity of mercury (ρ) vs. temperature [96]: $\rho_\parallel$ and $\rho_\perp$ are resistivities in the direction of the trigonal axis and perpendicular to it.

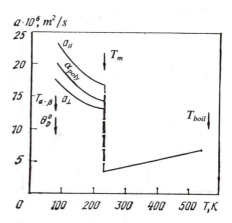

Figure 45 The diffusivity of mercury (a) vs. temperature [39]: $a_\parallel$ and $a_\perp$ are the values for the directions parallel to the trigonal axis and perpendicular to it, respectively.

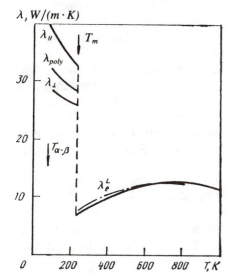

Figure 46 The thermal conductivity of mercury (λ) vs. temperature: $\lambda_\parallel$ and $\lambda_\perp$ are the respective values for the directions parallel and perpendicular to the trigonal axis [48].

§5 ALUMINUM, GALLIUM, INDIUM, THALLIUM

ALUMINUM

At standard pressure aluminum has an fcc structure with the parameter $a = 0.40496$ nm at 298 K. It contains three conduction electrons per atom, and therefore in

Table 17 The kinetic properties of mercury [39, 48, 96, 99–101]

T, K	$\rho \cdot 10^8$, $\Omega \cdot$ m		$a \cdot 10^7$, m²/s		λ, V/(m·K)	
	$\rho_\perp / \rho_\parallel$	ρ_{poly}	$a_\perp / a_\parallel$	α_{poly}	$\lambda_\perp / \lambda_\parallel$	λ_{poly}
100	8.60/6.48	7.89	16.7/22.0	19.0	28.6/39.4	32.0
200	19.71/14.82	18.08	13.6/17.5	15.0	25.8/34.1	29.0
300	—	102	—	4.2	—	8.0
400	--	113	—	5.4	—	10.0
500	—	126	—	6.2	—	11.0
600	—	137	—	—	—	12.0
700	—	150	—	—	—	13.0
800	—	160	—	—	—	13.0
900	—	—	—	—	—	12.5
1000	—	—	—	—	—	11.8

the free-electron approximation, the first Brillouin zone is all full. The second zone is nearly all full, but the Fermi surface does not touch the Brillouin zone boundary. In the third zone, the Fermi surface contains sets of arms [41]. Finally, the fourth zone contains a few very small isolated pockets of electrons.

Figure 47 shows the temperature dependence of the elastic constants. Above 150 K the constants c_{11} and c_{44} depend linearly on temperature. A small deviation from this linearity is observed only near the melting point. The elastic properties of high purity aluminum (99.999%) were investigated in [58]. The dependences of c_{11} and c_{44} agree with [97] to within 1–2%. They are, however, different for c_{12}. Above 300 K the values of c_{12} decrease with the increase in temperature. The behavior of the shear modulus c' of aluminum is typical for the fcc metals. The values of c' change more rapidly with temperature than the value of c_{44}. The coefficient of anisotropy $A = 2c_{44}/(c_{11} - c_{12})$ increases with temperature. It increases especially sharply above 200 K, when c' starts decreasing more rapidly than c_{44}.

At room temperature, polycrystalline aluminum has the following parameter values: $E = 0.71 \cdot 10^{11}$ Pa; $G = 0.265 \cdot 10^{11}$ Pa; $K = 0.74 \cdot 10^{11}$ Pa.

The temperature dependence of the electrical resistivity of aluminum has a long linear section (approximately from 150 to 600 K). On approaching the melting point, the value of $\partial \rho / \partial T$ starts increasing (Fig. 48). It should be pointed out

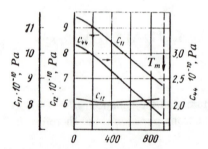

Figure 47 The elastic constants of aluminum vs. temperature [97].

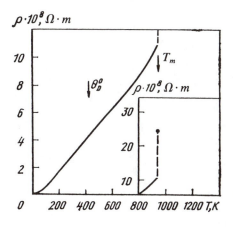

$\rho \cdot 10^8, \Omega \cdot m$

Figure 48 The average temperature dependence of the electrical resistivity of aluminum (ρ) [99, 100].

that, at moderate temperatures, the resistivity values for aluminum, quoted in reference books, differ by almost 10% [99–101].

The absolute values of diffusivity of aluminum are high. The diffusivity has a negative temperature coefficient in the solid state at $T > 150$ K, and a positive coefficient in the liquid state. The data shown in Fig. 49 refer to a sample with the relative residual resistance of $r = 1600$. The uncertainty in the data is 4% within the 700–900 K temperature interval, and 8% outside of it [39].

For the solid and the liquid states, the high values of the thermal conductivity of aluminum are due to its electronic component. Figure 50 shows the data for a sample with $r = 1600$. Calculations prove that at 800 K the fraction of the lattice contribution does not exceed 2%. This is less than the uncertainty of the given data which is approximately 5% [48].

The absolute thermoelectric power of aluminum is negative and depends on temperature in a complex manner (Fig. 51) [55, 102].

In the solid state, at $T > 400$ K, the Hall coefficient of aluminum is equal to $3.9 \cdot 10^{-11}$ m^3/C, and remains almost the same during melting [113].

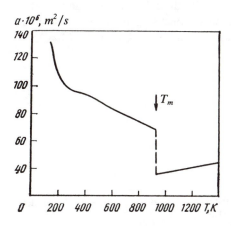

$a \cdot 10^6, m^2/s$

Figure 49 The diffusivity of aluminum (a) vs. temperature [39].

$\lambda, W/(m \cdot K)$

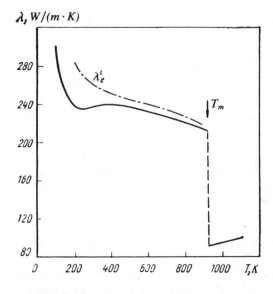

Figure 50 The thermal conductivity of aluminum (λ) vs. temperature [48].

GALLIUM

At standard pressure and 297 K, gallium has an orthorhombic structure with the parameters $a = 0.45197$, $b = 0.45260$, $c = 0.76633$ nm. Gallium has a low melting point ($T_m = 302.94$ K) [74]. The Debye temperature, which was determined by the calorimetric measurements at low temperatures, is $\theta_D^0 = 317$ K, which exceeds the melting point. Therefore the 'low-temperature' region covers, partially, the liquid state region for this metal.

The orthorhombic unit cell of gallium contains eight atoms, making its Fermi surface very complex [41].

The presence of parts of the Fermi surface in a large number of bands causes difficulties for the accurate determination of their sizes and parameters, as they are very sensitive to even insignificant changes in the Fermi level.

Figure 52 shows the elastic constants of gallium as functions of temperature from 80 to 293 K [103]. The temperature dependences c_{ij} are almost linear, although some nonlinearity is observed for c_{12} and c_{31} (for which $\partial c/\partial T > 0$).

$S \cdot 10^6, V/K$

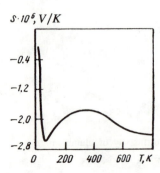

Figure 51 The thermoelectric power of aluminum (S) vs. temperature [55, 101].

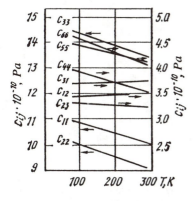

Figure 52 The elastic constants of gallium vs. temperature [103].

The anisotropy in the crystalline structure and, consequently, the strong anisotropy of the Fermi surface result in a substantial anisotropy in the kinetic properties of gallium. Thus, at room temperatures, the anisotropy in its electrical conductivity ρ_c/ρ_b reaches 6.8 [104], which, for metals, is the highest value of the resistance anisotropy. An interesting feature of the data shown in gallium (S) is -1 $\mu V/K$ near the melting point, and during melting it increases to 0.5 $\mu V/K$ [113]. In the solid state, the Hall coefficient is $R = 3.9 \cdot 10^{-11}$ m^3/C, and it decreases during melting by 5%.

INDIUM

At standard pressure, indium has a bct structure with the parameters $a = 0.32512$, $c = 0.49467$ nm at 293 K [74]. This structure makes the Fermi surface of indium resemble that of aluminum [41].

Table 18 gives the elastic constants of indium up to the melting point [107, 108]. The shear elastic constant $c' = (c_{11} - c_{12})/2$ decreases very sharply, by

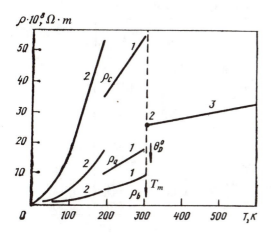

Figure 53 The temperature dependence of the electrical resistivity of single crystal gallium (ρ) in the directions of the a-, b-, and c-axes (ρ_a, ρ_b, ρ_c): 1) [104]; 2) [113].

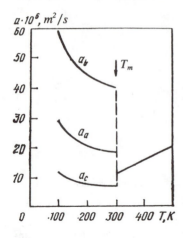

$a \cdot 10^6$, m^2/s

Figure 54 The diffusivity of gallium (a) in the directions of the a-, b-, and c-axes vs. temperature [39].

almost 90%, in the interval from 4 K to the melting point. As a result, the temperature dependence of the elastic anisotropy of indium is non-trivial. For indium, four coefficients of the elastic constants $A = c_{44}/c_{66}$, $B = c_{33}/c_{11}$, $C = c_{12}/c_{13}$, and, $D = 2c_{66}/(c_{11} - c_{12})$ describe the Fig. 53 is that during melting the electrical resistivity decreases in the direction of the c-axis and increases in the directions of the a- and b-axes [104, 106]. The diffusivity of gallium is characterized by a similar anisotropy (Fig. 54). The uncertainty in the given values of a is $\approx 7\%$ for the solid state and 15% for the liquid state. It should be pointed out that the diffusivity values in the direction of the a-axis resemble the data for a polycrystalline sample [39].

Similar to the absolute values of λ, the anisotropy in the thermal conductivity of gallium is due to an electronic component (Fig. 55), especially in the b- and a-directions. The uncertainty of the given data is 4–10% for the solid state and about 15% for the liquid state [39, 105].

Table 18 The elastic constants of indium ($c_{ij} \cdot 10^{-10}$, Pa)

T, K	c_{11}	c_{12}	c_{13}	c_{33}	c_{44}	c_{66}
			According to data of [107]			
4.2	5.392	3.871	4.513	5.162	0.797	—
77	5.260	4.056	4.457	5.080	0.764	—
300	4.535	4.006	4.151	4.515	0.651	—
			According to data of [108]			
300	4.51	3.97	4.11	4.53	0.653	1.19
351	4.33	3.98	4.09	4.35	0.630	1.07
371	4.25	3.96	4.04	4.29	0.617	1.03
391	4.21	3.95	4.03	4.23	0.603	0.988
422	4.11	3.92	3.99	4.13	0.583	0.921
429.7	4.09	3.91	3.95	4.11	0.578	0.905

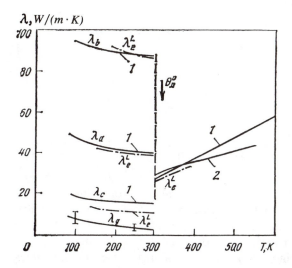

Figure 55 The thermal conductivity of gallium (λ) in the directions of the a-, b-, and c-axes vs. temperature: *1*) [48]; 2) [105]; λ_g is calculated in [23]; λ_e^L is calculated using the W-F-L law.

The thermoelectric power of solid polycrystalline has a degree of elastic anisotropy. The temperature dependences of these coefficients are shown in Fig. 56.

The temperature dependence of the electrical resistivity of indium is shown in Fig. 57. The data describe a high purity sample (99.999%) with an uncertainty of about 5%. The temperature dependence of the condensed indium is typical for nontransition metals ($\partial^2\rho/\partial T^2 > 0$).

According to the data given below, the temperature dependence of the diffusivity of indium also does not have any unusual features [39, 48]

T,K	100	200	300	400	500	600	700
$a \cdot 10^o$, m^2/s	64.6	54.0	47.9	40.8	22.2*	24.1*	26.2*
λ, W/(m·K)	97.6	89.7	81.7	74.5	37.2*	38.9*	—

*Preliminary data.

The listed data refer to a sample with $r \approx 15,000$; the uncertainty is estimated as 8% for the solid and 15% for the liquid states.

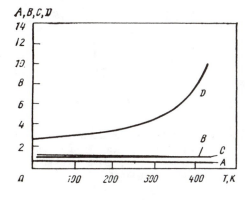

Figure 56 The coefficients of the elastic anisotropy of indium (A, B, C, D) vs. temperature [108].

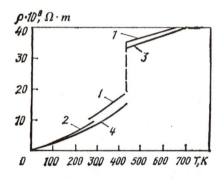

Figure 57 The electrical resistivity of indium (ρ) vs. temperature: *1*) [109]; *2*) [104]; *3*) [113]; *4*) [105].

In the solid and liquid states, the thermal conductivity of indium is of an electronic nature, as in the previously discussed metals. It should be mentioned that $\lambda \approx \lambda_e$ is within an uncertainty of 10% for the solid state and 15–20% for the liquid state. The data, quoted above and shown in Fig. 58, apply to high purity indium (99.999%). It is interesting to note that the Lorenz number of indium does not change significantly during melting [110].

The information on the thermoelectric power and Hall coefficient of indium is given in Fig. 59 [109, 92, 55, 112].

THALLIUM

Thallium, at standard pressure, has two polymorphic modifications. At low temperatures, an hcp modification is stable. It has the parameters $a = 0.34566$ nm and $c = 0.55248$ nm at 291 K. A bcc modification is stable from 507 K to the melting point ($T_m = 577$ K [74] or 576.2 K [39]) and has the parameter $a = 0.3882$ nm.

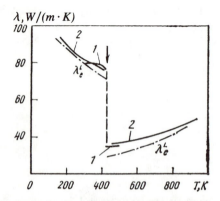

Figure 58 The thermal conductivity of indium (λ) vs. temperature: *1*) [110]; *2*) [48].

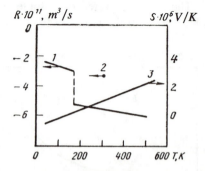

Figure 59 The absolute thermoelectric power (*S*) and Hall coefficient (*R*) of indium vs. temperature: *1*) [109]; *2*) [92]; *3*) a combined curve for In, Ga, Tl [55].

The Fermi surface of thallium consists of six sheets distributed among the third and sixth Brillouin zones [41]. Its main sheets are: a large closed surface in the third zone; a smaller closed sheet of holes in the same zone; a honeycomb-like network of surfaces in the fourth zone; and small pockets of electrons in the fourth, fifth and sixth zones. The elastic constants of thallium ($c_{ij} \cdot 10^{-10}$, Pa) are given below [111].

T, K	4.2	50	75	100	150	200	300
c_{11}	4.440	4.395	4.350	4.300	4.215	4.160	4.080
c_{33}	6.020	5.940	5.885	5.820	5.700	5.565	5.280
c_{44}	0.880	0.859	0.837	0.810	0.762	0.743	0.726
c_{66}	0.341	0.331	0.325	0.319	0.307	0.295	0.270
c_{12}	3.76	3.73	8.70	3.66	3.60	3.55	3.54
c_{13}	3.00	3.00	3.00	2.9	2.8	2.8	2.9

Although in the hexagonal modification the ratio of the lattice parameters $c/a = 1.59$ is close to the ideal value (1.63), the degree of the elastic anisotropy for thallium is unusually high. In the temperature interval 4.2–300 K, the coefficients, defining the anisotropy of the hexagonal lattice, vary from $2c_{44}/(c_{11} - c_{12}) = 2.56$ at 4.2 K to 2.70 at 300 K; and from $c_{11}/c_{33} = 0.74$ at 4.2 K to 0.78 at 300 K.

The kinetic properties of thallium have not been sufficiently studied. The preliminary data on the electrical resistivity of polycrystalline samples (Fig. 60) showed the temperature dependence to be about the same as for other metals of this subgroup, i.e. $\partial^2 \rho / \partial T^2 > 0$ for the solid state, and a linear increase in the resistance for the liquid state.

The diffusivity and thermal conductivity of polycrystalline thallium are listed below [39, 40].

T, K	100	200	300	400	500
$a \cdot 10^{6}$, m²/s	38.7	32.8	30.1	27.6*	25.2*
λ, W/(m·K)	55.6	49.4	45.1	43.8*	42.1*

*The data need further investigation.

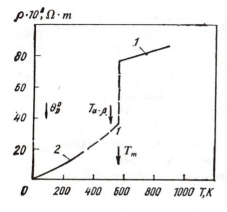

$\rho \cdot 10^{8}$, $\Omega \cdot m$

Figure 60 The electrical resistivity of polycrystalline thallium (ρ) vs. temperature: *1*) [105]; 2) [82].

The diffusivity data have an uncertainty of 15%. In this case, as well, the temperature dependence $\partial a/\partial T < 0$ was observed, which is typical for the metals of this subgroup.

The thermal conductivity of polycrystalline thallium, like other ordinary metals, is of an electronic nature, $\lambda/\lambda_e = 1.1$ at 100 K and 1.04 at 300 K. The difference between λ and λ_e is within the uncertainty in the determination of λ, which is about 10%. The thermoelectric power of thallium (S) is 0.4 $\mu V/K$ in the solid state, and 0.5 $\mu V/K$ in the liquid state (near the melting point). The corresponding values of the Hall coefficient are $R = 6.27 \cdot 10^{-11}$ and $4.76 \cdot 10^{-11}$ m^3/C [112, 113].

§6 TIN, LEAD

TIN

At standard pressure, tin has two crystalline modifications. β-tin is stable above 286.2 K, and has a bct structure with the parameters $a = 0.58317$, $c = 0.31813$ nm at 298 K. The low-temperature modification, α-tin, has an fcc structure with the parameter $a = 0.64892$ nm at 298 K [39, 74]. Only β-tin has a metal-like conductivity.

The unit cell of β-tin contains two atoms and eight conduction electrons. Therefore, as a minimum, tin should have four filled zones [41]. The free-electron model correctly describes the Fermi surface of β-tin. The first, second and third Brillouin zones are very nearly full, except for the small pockets of holes at the W point in the second zone and a multiply-connected surface of holes in the third zone. The fourth zone has a closed surface of electrons, centered at the Γ point, surrounded by a multiply-connected surface of holes. The fifth zone has an isolated electronic sheet around the Γ point, a multiply-connected surface of electrons consisting of 'pears' centered at the H point, and 'two stuck pancakes' around the V point. Finally, the sixth zone has two isolated electronic sheets (Fig. 61) [41].

A simplest approximation for the Fermi surface of k-tin is a spherical surface overlapping box-like boundries (Fig. 62). The surface area, projected on the plane parallel to the X-axis (A_3), is about 1.5 times smaller than the area projected in the perpendicular direction (A_1).

Table 19 gives the elastic constants of tetragonal tin for the interval from 4.2 K to the melting point [114]. The constants change monotonically with temperature; the values of c_{33}, however, remain almost constant. The appreciable temperature changes in the shear modulus $c' = (c_{11} - c_{12})/2$ can be explained by the existence of pockets of electrons and holes on the Fermi surface. These changes indicate that the anisotropy of elasticity increases with temperature [114]. Figure 63 shows, up to the melting point, the temperature dependence of Young's modulus for polycrystalline tin [115].

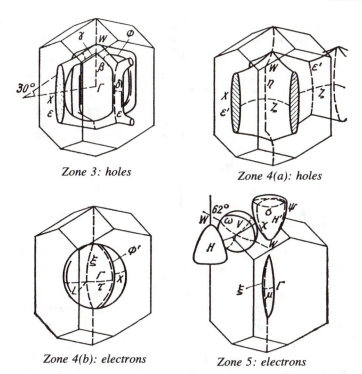

Zone 3: holes

Zone 4(a): holes

Zone 4(b): electrons

Zone 5: electrons

Figure 61 The Fermi surface of tin as an approximated surface of nearly-free electrons [41].

The temperature dependence of the electrical resistivity of β-tin is shown in Fig. 64. Below 286.2 K this modification is metastable, and therefore the kinetic parameters may be time-dependent. The anisotropy in the electrical resistivity was studied extensively in [116, 117] (Fig. 64). The ratio $\rho_\parallel / \rho_\perp$ reaches a maximum at about 20 K, and then approaches a constant value of $a_\infty = \rho_\parallel / \rho_\perp = 1.48$.

In a tetragonal crystal an expression can be written for any direction of the current: $\rho(\varphi) = \rho_\perp [1 + (a_\rho - 1)\cos^2\varphi]$, where φ is the angle between the tetragonal axis and the direction of the current ($\rho_\perp = \rho$ at $\varphi = 90°$); $a_\rho = \rho(\varphi =$

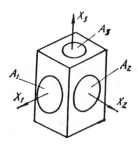

Figure 62 The simplest model for the Fermi surface of tin [92].

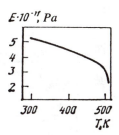

Figure 63 Young's modulus (E) of tin [115] vs. temperature.

Table 19 The elastic constants of tin ($c_{ij} \cdot 10^{-10}$, Pa)

T, K	c_{11}	c_{12}	c_{13}	c_{33}	c_{44}	c_{66}
According to data of Rain from [114]						
4.2	8,274	5.785	3.421	10.310	2.695	2.818
77	8.152	5.790	3.642	10.040	2.620	2.781
300	7.230	5.940	3.578	8.840	2.203	2.400
According to data of [114]						
301	7.20	5.85	3.74	8.80	2.19	2.40
418	6.58	5.86	3.77	8.08	1.91	2.14
459	6.36	5.85	3.78	7.84	1.81	2.03
501	6.14	5.83	3.79	7.58	1.70	1.91
505	6.13	5.85	3.80	7.56	1.69	1.90

$0°)/[\rho(\varphi = 90°)] = \rho_\parallel/\rho_\perp$. The analysis of this dependence [92] showed that, for the lowest temperatures, it is determined by the ratio of the impurity components. In regions of the maximum values of a_ρ, this dependence corresponds approximately to a_∞^2, where a_∞ is the value at $T \to \infty$. At intermediate temperatures, the value of a_ρ is proportional to T^{-2} and, finally, as $T \to \infty$, it approaches a constant value determined by the projection ratio of the corresponding parts of the Fermi surface $\rho_\parallel/\rho_\perp = A_\perp/A_\parallel$. The presence of a maximum in this curve is explained by the anisotropic nature of the relaxation time and by the acute angles of the low

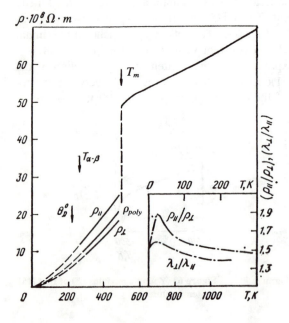

Figure 64 The temperature dependence of the electrical resistivity of tin (ρ): $\rho_\parallel$ and $\rho_\perp$ are the resistivity values in the direction of the tetragonal axis and perpendicular to it. In the inset, the temperature dependences of the anisotropy in the electrical resistivity ($\rho_\parallel/\rho_\perp$) and thermal conductivity ($\lambda_\perp/\lambda_\parallel$) [116–118].

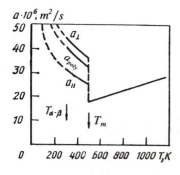

Figure 65 The temperature dependence of the diffusivity of tin (a) in the direction of the tetragonal axis ($a_\parallel$), perpendicular to it ($a_\perp$), and also for a polycrystalline sample (a_{poly}) [48, 119].

temperature electron-phonon scattering $a_s = (A_\perp/A_\parallel) \cdot \tau_\parallel/\tau_\perp$, where $\tau_\parallel$ and $\tau_\perp$ are the relaxation time for the corresponding directions.

The acute angle of scattering causes the Brownian motion of the scattered electrons through the corresponding parts of the Fermi surface, and near the maximum $\tau_\parallel/\tau_\perp = A_\perp/A_\parallel$, which leads to $\rho_\parallel/\rho_\perp = (A_\perp/A_\parallel)^2$. With increasing temperature the average angle of scattering increases, the relaxation time becomes more isotropic, and, in the region of intermediate temperatures, the anisotropy decreases and assimptotically approaches $\rho_\parallel/\rho_\perp = A_\perp/A_\parallel$. This analysis does not take into account the anisotropy in the phonon spectrum, the electron-phonon scattering, and the fine details of the Fermi surface. Nevertheless, as a first approximation, it describes well the experimental data. In the liquid state, the temperature dependence of the electrical resistivity of tin is close to linear: $\rho = 63.889 + 0.024515 (T - 1173)$ ($\mu\Omega \cdot$ cm), where $1173 < T < 2200$ K [118]. According to [105], $\rho = 34.16 + 2.57 \cdot 10^{-2}T$ for the interval 505–1280 K.

Figure 65 gives the temperature dependence of the diffusivity for solid and liquid tin. In the liquid state, the diffusivity (in $a \cdot 10^6$, m^2/s) is described by the equation $a = 30.626 + 1.3016 (T - 1273)$, where $T < 2100$ K [118].

Figure 66 shows the temperature dependences of the thermal conductivity of solid and liquid tin. The temperature dependence of the anisotropy in the thermal conductivity of β-tin resembles that of the electrical resistivity; the maximum value of $\lambda_\perp/\lambda_\parallel$, however, is lower (see Fig. 66). The same scattering mechanism is

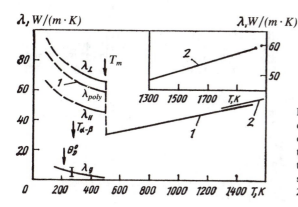

Figure 66 The temperature dependence of the thermal conductivity (λ) of single crystal tin in the direction of the tetragonal axis ($\lambda_\parallel$), perpendicular to it ($\lambda_\perp$), and for a polycrystalline sample (λ_{poly}) [48, 118, 119]: *1*) [48]; *2*) [118].

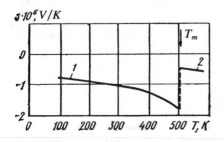

Figure 67 The absolute thermoelectric power (S) of tin vs. temperature: _1_) [55]; 2) at $T > T_m$ [113].

responsible for changes in the anisotropies of the thermal conductivity and electrical resistivity. Therefore, at high temperatures, these anisotropies are equal within the uncertainty of $\approx 10\%$.

It should be added that the thermal conductivity of β-tin is of the electronic nature, and above 200 K $\lambda \approx \lambda_e$.

The thermoelectric power of tin is given in Fig. 67. The thermoelectric power is negative, and for the solid state its absolute value increases with temperature. As a first approximation, this can be explained by the predominantly electronic nature of the Fermi surface. The value of the Hall coefficient of tin is close to R_0, i.e. to the value calculated using the free-electron model [113]; $R_0 = 4.42 \cdot 10^{-11}$ m^3/C; $R_{liq}/R \approx 1.0$.

LEAD

At standard pressure, lead has an fcc structure with the parameter $a = 0.49502$ nm at 298 K. The free-electron model provides a good description of the Fermi surface of lead [41]. The first Brillouin zone is completely full, the second zone consists of a distorted sphere which encloses a region of holes centered at the Γ point. The third zone appears as a multiply-connected set of tubes. Finally, the fourth zone contains six isolated pockets of electrons at the corners of the Brillouin zone [41]. The elastic constants of lead ($c_{ij} \cdot 10^{-10}$, Pa) are given below [108, 120].

T, K	0	80	160	240	300	418	501	540	586	600.5
c_{11}	5.554	5.429	5.261	5.083	4.953	4.682	4.466	4.351	4.220	4.150
c_{12}	4.542	4.483	4.395	4.297	4.229	4.066	3.934	3.863	3.782	3.760
c_{44}	1.942	1.836	1.709	1.583	1.490	1.312	1.187	1.124	1.053	1.023

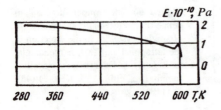

Figure 68 Young's modulus (E) of lead vs. temperature [115].

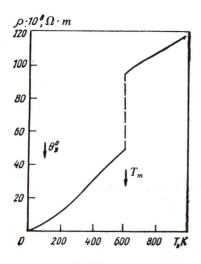

$\rho \cdot 10^9, \Omega \cdot m$

Figure 69 The electrical resistivity (ρ) of lead vs. temperature [105, 121, 122].

Below 300 K the above table quotes the data of [120]; above 300 K, the data of [108]. The elastic anisotropy of lead increases with temperature, especially rapidly near the melting point. Young's modulus of polycrystalline lead increases near the melting point; after melting, it decreases rapidly (Fig. 68).

Figure 69 shows the temperature dependence of the electrical resistivity of lead of 99.99% purity. In the temperature interval of 260–550 K, the electrical resistivity of high purity lead (ρ, $\mu\Omega \cdot cm$) ($r = 980$) is described by the equation $\rho = -0.9102 + 7.0943 \cdot 10^{-2} \, T + 3.326 \cdot 10^{-6} \, T^2 + 2.354 \cdot 10^{-8} \, T^3$. The uncertainty in the measurements is 0.4%. For the liquid state the resistivity is $\rho = 65.7 + 4.65 \cdot 10^{-2} T$ ($600 < T < 1200$ K)

Figure 70 shows the diffusivity of lead. It has a negative temperature slope in the solid state and positive slope in the liquid state. The uncertainty in the data given in [39] is 5% at moderate temperatures; about 8% near the melting point for the solid state; and, about 15% for the liquid. All of the data apply to the high purity (99.99%) metal.

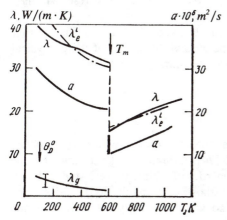

$\lambda, W/(m \cdot K)$ $a \cdot 10^6, m^2/s$

Figure 70 The thermal conductivity (λ) and diffusivity (a) of lead vs. temperature [122, 48, 39]; λ_g is the calculation of [23].

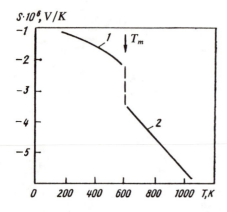

$S \cdot 10^6, V/K$

Figure 71 The absolute thermoelectric power of lead (S) vs. temperature: *1*) [55]; 2) [113].

As in the case of tin, the thermal conductivity of lead (see Fig. 70) is of an electronic nature in the liquid and solid states, and above 200 K $\lambda_e^L \approx \lambda_e^g$. The temperature dependence of the absolute thermoelectric power is given in Fig. 71.

In the solid state, near the melting point, the Hall coefficient of lead is equal to $R = 5 \cdot 10^{-11}$ m^3/C. During melting, it decreases by 10%.

§7 ARSENIC, ANTIMONY, BISMUTH

ARSENIC

At standard pressure, arsenic has a rhombohedral structure with the parameters a = 0.41318, α = 54.13° at 296 K [74]. Arsenic belongs to the semimetals rather than to the metals [41]. The electronic properties of arsenic are poorly known. The main feature of its Fermi surface, which is similar to that of bismuth and antimony, is the presence of small pockets of carriers [41]. Impurities considerably change the kinetic properties of arsenic. At room temperature, arsenic has the following physical properties: $E = 7.75 \cdot 10^9$ Pa; the electrical resistivity is ρ = 33.3 $\mu\Omega \cdot$ cm; the temperature coefficient of resistivity is $\partial\rho/\rho\partial t = 3.9 \cdot 10^{-3}$ K^{-1}, $a = 26.5 \cdot 10^{-6}$ m^2/s, λ = 35.6 W/(m $\cdot$ K); and, the Hall coefficient is R = 450 $\cdot 10^{-10}$ m^3/C [39, 48, 82]. The temperature dependences of these physical constants are assumed to be similar to those of antimony and bismuth.

ANTIMONY

At standard pressure, antimony has a rhombohedral structure with the parameters a = 0.45067 nm, α = 57.107° at 298 K [74]. Antimony, like arsenic, can be considered a semimetal. The electronic properties of antimony were studied better than those of arsenic, but much less than those of bismuth. Its Fermi surface resembles that of bismuth and has small pockets of holes in the fifth zone and electrons in the sixth zone [41]. Impurities considerably change its electronic characteristics. The success in studying its electric properties was achieved largely with recently obtained high purity samples (less than 0.01−0.001% of impurities).

From 300 to 873 K the elastic constants depend linearly on temperature [123]. The constants are described by the following equations: $c_{11} = (117 - 47.5 \cdot 10^{-3} T)$; $c_{44} = 42.4 - 9.3 \cdot 10^{-3} T$; $c_{33} = 51.2 - 21.6 \cdot 10^{-3} T$; $c_{12} = 31.6 - 14.3 \cdot 10^{-3} T$; $c_{13} = 27.6 - 9.8 \cdot 10^{-3} T$; $c_{14} = 26.2 - 7.0 \cdot 10^{-3} T$ $(c_{ij} \cdot 10^{-9}$, Pa). The uncertainty in the determination of the constants c_{11}, c_{13}, c_{33} is 8−12%, and 25−30% for all other constants.

Figure 72 gives the temperature dependence of Young's modulus for single crystal antimony of 99.999% purity for the three main crystallographic directions. According to [124], the additional decrease in the modulus for the [$\bar{1}$10] and [11$\bar{2}$] directions is caused by a redistribution of electrons and holes in small pockets, the stresses in these directions deforming the Fermi surface differently. The stress in the direction [111] produces an almost equal effect on all pockets, and a decrease in Young's modulus is not observed [124].

The temperature dependence of the electrical resistivity of antimony is shown in Fig. 73. Since the concentration of the free carriers is small, the resistivity is rather high at all temperatures, and near the melting point it reaches 140 $\mu\Omega \cdot$ cm, i.e. the value which is typical for metals with a short mean free path of electrons. The electrical resistance decreases by 30% during melting.

Table 20 gives the information on the diffusivity, thermal conductivity, thermoelectric power and Hall coefficient of antimony. The high values of the ther-

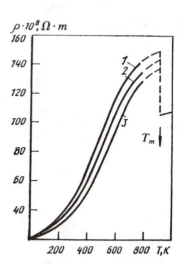

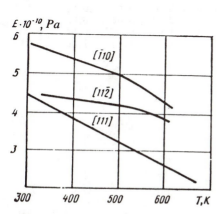

Figure 72 Young's modulus (E) of antimony vs. temperature [124].

Figure 73 The electrical resistivity of single crystal antimony (ρ) vs. temperature: *1*) $\rho_{\parallel}$, the axis of the sample is parallel to the bisectional axis of the crystal; *2*) ρ_{poly}; *3*) ρ_{33}, the axis of the sample is parallel to the trigonal axis of the crystal [125].

Table 20 The kinetic properties of antimony [39, 48, 113, 124, 125]

T, K	$a_{poly} \cdot 10^6$, m²/s	λ_{poly} W/(m·K)	$S \cdot 10^6$, V/K			$R \cdot 10^{18}$, m³/c		
			S_{11}	S_{poly}	S_{33}	R_{123}	R_{poly}	R_{331}
100	40.8	46.4	15		6			
200	22.5	30.2	33	43	17	22	21	20
300	17.5	24.3	50	45	27	26	24	22
400	15.0	21.2	56	47	30	22	20	18
500	13.4	19.4	53	44	28	16	15	14
600	12.3	18.2	48	40	24	12	11	10
700	11.5	17.4	40	35	20	9	8	7
800	10.8	16.8	33	27	15	7	6	5
1000	15.0	27.0	—	5.2	—	—	−0.44	—

Note: S_{11} and S_{33} are in the direction of the bisectional and trigonal axes, respectively.

moelectric power and Hall coefficient of antimony are typical for semimetals. The strong temperature dependence of the Hall coefficient indicates the development, with temperature, of activation processes of carriers.

BISMUTH

At standard pressure, bismuth, like arsenic and antimony, has a rhombohedral structure with the parameters $a = 0.4746$ nm, $\alpha = 57.23°$ at 298 K [74]. The structure of the Fermi surface of bismuth is well established [41]. The elastic constants of bismuth (c_{ij}, Pa) are listed below [114].

T, K	4.2	80	300	301	412	452	502	544
c_{11}	6.87	6.86	6.37	6.35	6.03	5.91	5.75	5.61
c_{12}	2.37	2.38	2.49	2.47	2.47	2.45	2.43	2.39
c_{13}	—	—	2.47	2.45	2.44	2.43	2.43	2.42
c_{14}	0.844	0.805	0.717	0.723	0.67	0.65	0.62	0.60
c_{33}	4.06	4.06	3.82	3.81	3.67	3.62	3.55	3.49
c_{44}	1.29	1.27	1.123	1.13	1.03	0.997	0.946	0.90
c_{66}	2.25	2.24	1.941	1.94	1.78	1.73	1.66	1.61

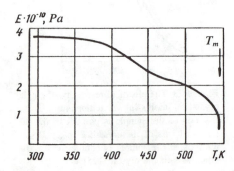

Figure 74 Young's modulus (E) of polycrystalline bismuth vs. temperature [115].

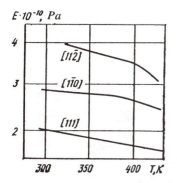

Figure 75 Young's modulus (*E*) of single crystal bismuth vs. temperature [126].

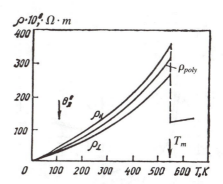

Figure 76 The temperature dependence of the electrical resistivity of bismuth (ρ) in the direction of the trigonal axis ($\rho_\parallel$), perpendicular to it ($\rho_\perp$), and for a polycrystalline sample (ρ_{poly}) [113, 127–129].

The temperature dependence of Young's modulus is shown in Fig. 74. The decrease above 400 K was probably caused by a redistribution of free carriers between the sheets of the Fermi surface [126]. This effect is also clearly visible in single crystals for the directions [110] and [112] (Fig. 75).

Figure 76 shows the temperature dependence of the electrical resistivity of high purity (99.999%) bismuth [127–131]. The anisotropy in resistivity reaches 30%, and, above 100 K, the resistivity has a weak temperature dependence. A number of studies, for example [129, 131], analyzed the scattering mechanisms and the kinetic properties of bismuth. At high temperatures, above $\theta_D^\circ = 119$ K, the anisotropy in resistivity is mainly determined by the anisotropy in the sheets of the Fermi surface. The electrical resistivity of bismuth reaches record high values ($\rho = 380$ $\mu\Omega \cdot$ cm near the melting point) because of a relatively small concentration of carriers. It is interesting that, during melting, the conductivity of bismuth increases by almost three times.

The diffusivity of bismuth has a negative temperature coefficient, and its anisotropy is similar to the anisotropy in resistivity (Fig. 77). The data, given above,

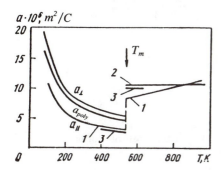

Figure 77 The temperature dependence of the diffusivity of bismuth (*a*) in the direction of the trigonal axis ($a_\parallel$), perpendicular to it ($a_\perp$), and for a polycrystalline sample (a_{poly}): *1*) [39]; *2*) [93]; *3*) [132].

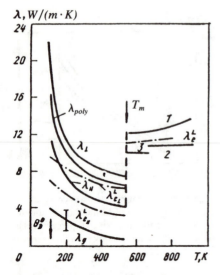

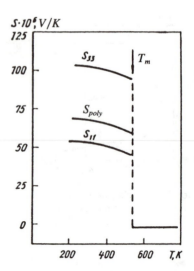

Figure 78 The thermal conductivity of bismuth (λ) vs. temperature, in the direction of the trigonal axis ($\lambda_\parallel$), perpendicular to it ($\lambda_\perp$), and for a polycrystalline sample (λ_{poly}): *1*) [48]; *2*) [51]; *3*) [132].

Figure 79 The absolute thermoelectric power (S) of bismuth vs. temperature [128].

describes high purity bismuth. The uncertainty is about 10% at room temperature, 15% above and below room temperature for the solid state, and about 20% for the liquid state [23]. The data for the diffusivity of liquid bismuth, however, require further investigation.

The thermal conductivity of bismuth is distinguished by large negative values above 100 K, and very low values above room temperature. Although the electronic component is the main factor that determines $\lambda(T)$ (Fig. 78), the lattice contribution is well pronounced at temperatures below 300 K. Within the uncertainty limits of λ ($\approx$10%), it can be stated that $\lambda - \lambda_e^L = \lambda_g$, where the lattice

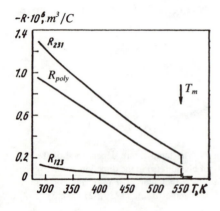

Figure 80 The Hall coefficient of bismuth (R) vs. temperature [128].

contribution is estimated using the method of [23]. In the liquid state, within the limits of 15%, $\lambda \approx \lambda_e$.

In the solid state, bismuth is characterized by record high values, for metals, of the absolute thermoelectric power (Fig. 79). These values lie between the characteristic values for typical metals and semiconductors. In the liquid state, the values of the thermoelectric power are small; their sign and magnitude are typical for metals with electronic conductivity [113].

The Hall coefficient of bismuth for the liquid state $R = -6.4 \cdot 10^{-10}$ m^3/C does not differ substantially from that, calculated using the nearly-free electron model. For the solid state, however, the Hall coefficient is highly anisotropic, and its absolute value and temperature coefficient are large (Fig. 80).

THREE

TRANSITION METALS

§1 SCANDIUM, YTTRIUM, LANTHANUM

SCANDIUM

Table 6 lists the main crystalline characteristics of scandium. The density of scandium is small. It is the lightest among the transition metals. The melting point of scandium is rather high. For temperatures below 1607 K, it has an hcp structure with, at room temperature, the lattice parameters $a = 0.30309$ nm, and $c = 0.52733$ nm. Above 1607 K scandium is transformed into a cubic modification with the lattice parameter $b = 0.4541$ nm.

Figure 81 shows the density of the electronic states for the $3d$-metals. The Fermi level of scandium is located near the maximum of the density of states [133].

The Fermi surface of scandium is completely contained in the third and fourth zones. It has a complex multiply-connected shape. The Fermi surface is distantly related to the free-electron model. The structure of the Fermi surface of scandium requires further investigation. It is assumed, however, to be similar to that of yttrium [41].

At $T = 298$ K, scandium has the following values for $c_{ij} \cdot 10^{-10}$, Pa [134]: $c_{11} = 9.93$; $c_{12} = 4.57$; $c_{13} = 2.94$; $c_{33} = 10.69$; $c_{44} = 2.77$; and, $c_{66} = 2.68$.

At room temperature, the values of the moduli of elasticity are: $E = 8.09 \cdot 10^{10}$, $G = 3.18 \cdot 10^{10}$, and $K = 5.84 \cdot 10^{10}$ Pa [135].

The information on the electrical resistivity of scandium is scarce (Fig. 82). In [136] the electrical resistivity was measured for polycrystals at low tempera-

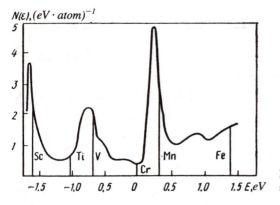

Figure 81 The density of the electronic states vs. energy for the 3d-metals [133].

tures, and in [137] for single crystals in the direction of the c-axis and perpendicular to it. At high temperatures, the electrical resistivity of polycrystalline samples was measured in [138, 139]. It should be mentioned that the investigated samples were not of high purity. For example, according to [138, 139], the relative residual resistivity was $r = 12$, and [138] investigated a sample of 99.8% purity. In addition, the data obtained for polycrystalline metals with anisotropic structures may be influenced by traces of texture, inherited after the thermomechanical treatment of samples [140]. In particular, the data of [139] agree with [137] for the perpendicular direction. For hexagonal metals the average value is $\rho_{av} = 1/3\rho_{\parallel} + 2/3\rho_{\perp}$. Below 1400 K this allows to obtain the probable values of resistivity. These values were obtained from [137], below 300 K, and from the average data of [138, 139], which agree with [137]. The uncertainty in these values is about 5–10%.

A distinguishing feature of the temperature dependence of $\rho(T)$ of scandium

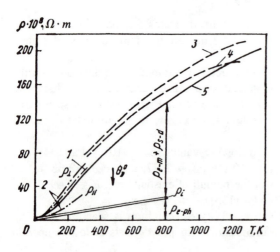

Figure 82 The electrical resistivity of scandium (ρ) vs. temperature for: *1*) a polycrystalline sample [136]; *2*) a single crystal [137]; *3*) a polycrystal [138]; *4*) a polycrystal [139]; *5*) the recommended average values.

is its negative curvature $(\partial^2\rho/\partial T^2 < 0)$ above 300 K. The main reason for this and the high values of the absolute resistivity is the electronic structure of scandium and, in particular, the high density of electronic states near the Fermi level. This high density is caused by the d-electrons of the shell forming a d-band, which can be hybridized with the p-states.

Figure 82 shows the estimated contributions to the electrical resistivity from different scattering mechanisms. The contribution of impurities was calculated using Matthiessen's rule $\rho_i = \rho_{0K} \approx \rho_{4.2K}$. The contribution of ρ_i to ρ decreases with the increase in temperature.

The electron-phonon contribution, caused by scattering of electrons by the vibrations of crystalline lattice, may be estimated through a comparison with calcium, the closest neighbour of scandium in the periodic table. The electron-phonon contribution to the resistivity of scandium can be expressed as

$$\rho_{e\text{-}ph}^{Sc} = \rho_{e\text{-}ph}^{Ca} M_{Ca} \theta_{D\ Sc}^2 / \left(M_{Sc} \theta_{D\ Ca}^2 \right) \tag{40}$$

where M_i and θ_{Di} are the corresponding atomic masses and Debye temperatures, respectively.

The results of these estimations are also shown in Fig. 82. It follows from this figure that, at 800 K, this contribution to the resistivity of scandium does not exceed 15%. The remaining part of the electrical resistivity $\Delta\rho = \rho - \rho_{e-ph} - \rho_i$ is due to the contributions that are specific to the transition metals.

Study [140] shows that scandium has two competing mechanisms of scattering. The mechanism of the band type is caused by the transitions of the s-electrons into the d-band. Another mechanism is due to the scattering of s-electrons by the magnetic inhomogeneities, particularly by paramagnons. The results of diffusivity measurements for scandium were given only in [140, 141], and, below 900 K, no direct measurements were performed. Study [39] contains some preliminary estimations. The data between 300 and 900 K in Fig. 83 were derived by extrapolation. Study [140] points out that texture and thermomechanical treatment considerably influence diffusivity.

The average kinetic properties of polycrystalline scandium are given below. The uncertainty is 5–10% [39, 48, 137–141].

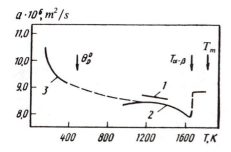

Figure 83 The diffusivity of polycrystalline scandium (a) vs. temperature: *1*) [141]; *2*) [140]; *3*) [39].

T, K	100	200	300	400	500	600	700
$\rho \cdot 10^8$, $\Omega \cdot$m . . .	12	30	52	72	91	109	124
$a \cdot 10^6$, m^2/s . . .	—	9,7	9,35	9,15	9,0	8,85	8,75
λ, W/(m·K) . . .	—	15,5	16,0	16,2	16,5	16,7	17,0

Continuation

T, K	800	900	1000	1200	1400	1600	1800
$\rho \cdot 10^8$, $\Omega \cdot$m . . .	138	151	163	186	204	—	—
$a \cdot 10^6$, m^2/s. . . .	8.65	8,60	8,50	8,40	8,30	7,95	8.85
λ, W/(m·K) . . .	17,2	17,5	17,7	18,2	18,8	19,2	—

The thermal conductivity of polycrystalline scandium for 300–1500 K is shown in Fig. 84, which combines the data for high [139, 140, 141] and low [48] temperatures. The uncertainty in the given data is about 10%. The lattice thermal conductivity was calculated in [140] using equations (21, 22) and taking into account the phonon-electron scattering.

Study [140] stated that the overestimation of the Lorenz number may be the result of band effects. However, calculations [140] made on the basis of the s-d scattering model yielded a value for L which was considerably greater than any previous value. It is possible that the scattering of electrons by paramagnons for which $L < L_0$, takes place in scandium. The combination of these two mechanisms probably yields the observed value of L.

The temperature dependence of the thermoelectric power of polycrystalline scandium has a maximum near 1400 K [55]. The positive sign of the thermoelectric power indicates that this value is largely influenced by the parts of the Fermi surface that contain holes. The presence of a maximum and a later decrease in the thermoelectric power indicate an increase in influence of the parts on the surface containing electrons (Fig. 85). According to [55], the overall uncertainty in the values of the thermoelectric power is about 10%.

Figure 86 shows the Hall coefficient of 99.9% pure polycrystalline scandium [143]. At 300 K, the Hall coefficient of single crystal scandium is $R_\perp = -0.957 \cdot 10^{-10}$ m^3/C, $R_{\parallel} = -0.305 \cdot 10^{-10}$ m^3/C [92, 144].

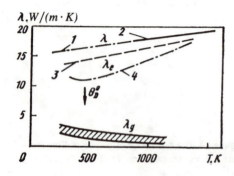

λ,W/(m·K)

Figure 84 The thermal conductivity of scandium (λ) vs. temperature: 1, 2) [139] and [140], respectively, polycrystalline samples; 3) the electronic component $\lambda_e^g = \lambda - \lambda_g$; 4) the same as for 3, $\lambda_e^L = L_0 T/\rho$; λ_g is the calculation from [23].

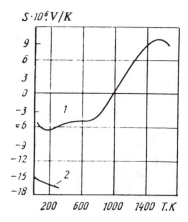

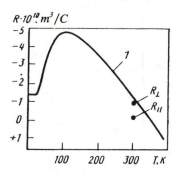

Figure 85 The thermoelectric power of scandium (S) vs. temperature: *1*) [55]; *2*) [142].

Figure 86 The Hall coefficient of scandium (R) vs. temperature: *1*) [143]; $R_\perp$ and $R_\parallel$ [92, 144].

YTTRIUM

Yttrium, like scandium, has an hcp structure at moderate and high temperatures with the lattice parameters $a = 0.36474$ nm, $c = 0.57306$ nm at 298 K. At 1753 K, the hcp structure is transformed into an fcc structure with the parameter $a = 0.411$ nm [39, 74]. The density states of yttrium was obtained by investigating the electronic heat capacity of alloys. Figure 87 shows that the Fermi level of yttrium is located near the maximum of $N(\varepsilon)$. The Fermi surface of yttrium is multiply-connected. Loucks calculated that it is entirely contained in the third and fourth Brillouin zones. It differs from the Fermi surface of scandium mainly by the fact that, at the M point, the fourth zone is empty.

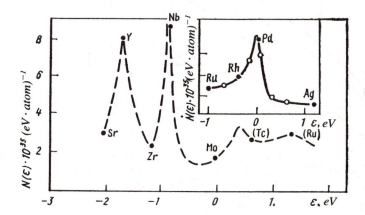

Figure 87 The density of states of the 4d-metals as a function of energy [133].

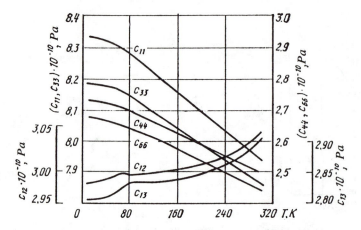

Figure 88 The elastic constants of yttrium vs. temperature [145].

Figure 88 shows the temperature dependence of the elastic constants of high purity yttrium. For investigated crystals the ratio of the electrical resistivity varied $r = 20 \div 420$ [145]. The data of [145] differ from an earlier investigation [146] by 2–3% with the exception of the elastic constant c_{13}. The elastic constants of yttrium $(c_{ij} \cdot 10^{-10}$, Pa) are given below [146].

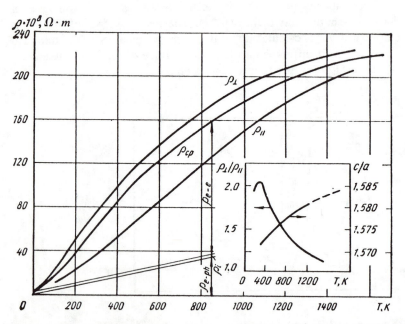

Figure 89 The electrical resistivity (ρ) [140], the anisotropy in it $(\rho_\perp/\rho_\parallel)$, and the ratio of lattice parameters (c/a) of yttrium vs. temperature.

T, K	300	350	400	c_{44}	2.431	2.382	2.333
c_{11}	7.79	7.71	7.63	c_{12}	2.85	2.85	2.84
c_{33}	7.69	7.66	7.64	c_{13}	2.10	2.10	2.10

The average electrical resistivity of polycrystalline and single crystal yttrium is assessed in [140] and shown in Fig. 89; $\rho_\parallel$ and $\rho_\perp$ are electrical resistivities in the direction of the c-axis and perpendicular to it, respectively. Yttrium has a large anisotropy of ρ: at 300 K $\rho_\perp/\rho_\parallel = 2.05$. The above data apply to samples with a residual resistance of 20–30 and a 99.0–99.9% purity. The data for polycrystalline yttrium were obtained from the relation $\rho_{av} = 2/3\rho_\perp + 1/3\rho_\parallel$; their uncertainty is 3–5%. Study [140] suggested that the high anisotropy of the kinetic characteristics demands measurements with either single crystals or polycrystalline samples with a strict texture control. This is because the thermomechanical history strongly influences the growth of orientated crystals.

Figure 89 shows that, at 800 K, the contributions from impurities and ordinary electron-phonon interactions amount to only about 25% of the total electrical resistivity of yttrium. For this calculation $\rho_i \approx \rho_{4.2K}$; the value of ρ_{e-ph} was determined from the electrical resistivity of strontium, a neighbour of yttrium in the periodic table. Two special features are evident. First, a strong non-linearity of $\rho(T)$ with $\partial^2\rho/\partial T^2 < 0$. Second, a decrease in the anisotropy with increase in temperature while the anisotropy of the lattice parameters increases and approaches the value of $c/a = 1.63$, (the 'ideal' for the hcp crystals).

Study [140] suggested that, in yttrium, the Mott s-d band mechanism may compete at moderate temperatures with that of the paramagnetic scattering of s-electrons. In addition, there may be a scattering in the local d-levels. The anisotropy in the electrical resistivity near the Debye temperature can be caused by the anisotropy of the Fermi surface. At high temperatures, the mean free path of electrons approaches the interatomic distance and this effect becomes substantial [140].

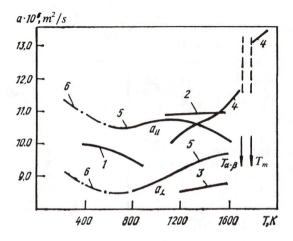

Figure 90 The diffusivity of yttrium vs. temperature: *1–4*) the data for polycrystals [39, 148–150], respectively; *5*) the data of [140] for a single crystal in the direction of the c-axis ($a_\parallel$) and perpendicular to it ($a_\perp$); *6*) the interpolated data of [140].

The diffusivity of single crystal yttrium ($r \approx 30$) was studied in [140] for the temperature range of 700–1600 K (Fig. 90). Figure 90 also shows the data for polycrystalline samples [147–150]. They correspond, approximately, to the results obtained for single crystals for the hexagonal direction and the direction perpendicular to it. This is probably caused by the texture of the samples. Table 21 summarizes the data for single crystal yttrium, with an uncertainty of about 5%. For temperatures below 700 K, the data were obtained through extrapolation, and the uncertainty is 10%.

The thermal conductivity data, given in [140] for the temperature interval 700–1600 K, apply to single crystal samples with $r \approx 30$. Below 500 K, the data for polycrystalline yttrium are either preliminary [48] or extrapolated [140]; the uncertainty is $\approx 15\%$ at 200 K (Fig. 91).

The lattice component of the thermal conductivity was derived by using equations (21) and (22) and taking into account the phonon-phonon and phonon-electron scattering processes. Above 700 K, the electronic component of the thermal conductivity, calculated by subtracting the lattice contribution from the total value, $\lambda_e = \lambda - \lambda_g$, is practically the same as the value computed from the W-F-L law. However, at temperatures below 700 K, a difference between these values may be observed due to the band and paramagnon effects [140].

The temperature dependence of the thermoelectric power of polycrystalline [151] and single crystal yttrium [15] is similar to that of scandium. At low temperatures, a minimum is observed; at moderate temperatures, the thermoelectric power changes in sign; and, at high temperatures, a maximum is observed (Fig. 92). The high temperature behavior of the thermoelectric power can be explained by the considerable decrease in the mean free path. Study [140] showed that if $\Lambda \to a$, then $\rho_\perp / \rho_\parallel \to 1$, $L \to L_0$, and S approaches the values typical for free electrons, which were in fact observed for yttrium and scandium.

TABLE 21 The kinetic properties of yttrium [39, 48, 140, 147]

T, K	$\rho \cdot 10^8$, $\Omega \cdot$ m			$a \cdot 10^4$, m²/s		λ, W/(m·K)	
	poly* calculated	$\parallel c$	$\perp c$	$\parallel c$	$\perp c$	$\parallel c$	$\perp c$
300	66.7	39.5	80.5	17.0	4.0	17.5	11.0
400	86.8	55.7	102.5	16.8	4.0	17.0	11.0
500	104.2	70.2	121.3	16.5	5.0	16.6	11.0
600	122.4	88.2	139.5	16.0	6.9	16.4	11.2
700	137.8	102.5	155.5	15.0	7.5	16.2	11.5
800	153.0	119.0	170.2	13.5	9.5	16.2	11.9
900	166.6	135.5	182.3	12.0	12.0	16.3	12.3
1000	179.1	150.5	193.5	10.0	15.0	16.5	13.1
1200	200.1	178.5	211.0	—	—	17.2	14.5
1400	213.6	203.0	219.0	—	—	17.7	16.0
1600	219.6	212.0	224.0	—	—	—	—

*poly: here and below means 'polycrystal'.

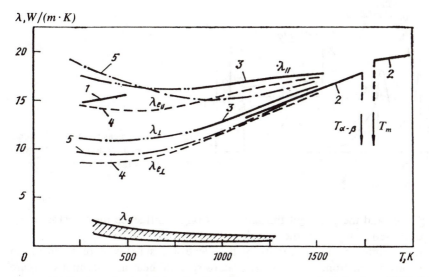

Figure 91 The thermal conductivity of yttrium (λ) vs. temperature: *1–3*) the data of [48], [147] and [140], respectively; *4*) $\lambda_e^g = \lambda - \lambda_g$; *5*) the calculation of λ_e^L from the W-F-L law; λ_g is the calculation by equations (34, 35).

The Hall coefficient, at room temperature, was given in [92]; $R_\perp = -0.47 \cdot 10^{-10}$ m³/C; $R_\parallel = -1.7 \cdot 10^{-10}$ m³/C.

LANTHANUM

Lanthanum has a more complex crystalline structure than yttrium and scandium: a double hcp structure below 583 K, a bcc structure between 583 and 1141 K, and an fcc structure from 1141 to the melting point at 1193 K [74]. Cooling and

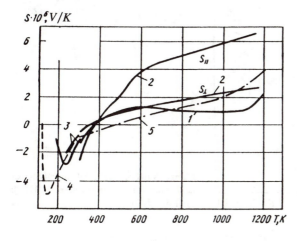

Figure 92 The absolute thermoelectric power (S) of yttrium vs. temperature: *1*) the data for a polycrystal [151]; *2, 3*) the data of [152, 153] for single crystals in the direction of the *c*-axis ($S_\parallel$) and perpendicular to it ($S_\perp$), respectively; *4, 5*) the data of Johansen and Miller [151].

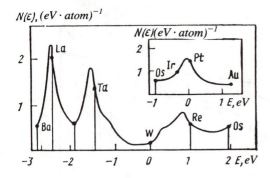

Figure 93 The density of the electronic states for the 5d-transition metals as a function of energy [133, 140].

heating rates and the purity of the samples strongly influence the transformation points and the properties of the samples [39, 74].

The information on the density of the electronic states was obtained, as in the previous cases, from the low-temperature specific heat investigations. Figure 93 shows that the Fermi energy of lanthanum lies near the maximum of the density of states [133, 140]. The Fermi surface of lanthanum is described in [41].

The information on the electrical resistivity of lanthanum is rather contradictory [140], which is probably caused by the low purity of the samples and by chemical reactions with the surrounding atmosphere.

The data of [140, 154, 155] are given in Fig. 94 for polycrystalline lanthanum of 99–99.5% purity and a relative residual resistance of $r \approx 10$. The data of [154] demonstrate the magnitude of the hysteresis in the electrical resistivity during heating and cooling in the region of the α-β transformation. At high temperatures, the data of [151] agree (within 10%) with the results of [155], which describe the γ and liquid phases.

The kinetic properties of lanthanum are listed below [39, 48, 140, 154, 155].

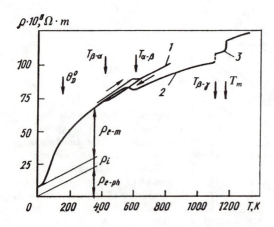

Figure 94 The electrical resistivity of lanthanum (ρ) vs. temperature: 1) [154]; 2) [140]; 3) [155].

T, K	100	200	300	400	500
$\rho \cdot 10^8$, $\Omega \cdot$m	25	50	61.7	72	79
$a \cdot 10^6$, m^2/s	9.2	10	10.9	11.7	12.7
λ, W/(m·K)	—	14.0	16.0	17.6	19.2

T, K	600	700	800	900	1000	1200
$\rho \cdot 10^8$, $\Omega \cdot$m	81	86	92	95	98	119
$a \cdot 10^6$, m^2/s	13.6	14.4	15.2	15.8	16.2	—
λ, W/(m·K)	20.8	22.4	23.6	24.8	26.0	—

Figure 94 shows the temperature dependence of the resistivity of lanthanum. The uncertainty in the resistivity values, given above and in Fig. 94, is about 10%.

The contribution of impurities was isolated as $\rho_i = \rho_{4.2K}$ (Fig. 94), and the electron-phonon contribution was calculated from the resistivity of barium. The residual part of resistivity $\Delta\rho = \rho - \rho_i - \rho_{e-ph}$ reaches saturation at 400–600 K. This is mainly due to the electron-electron scattering similar to the scattering by the magnetic inhomogeneities or the f-levels.

Figure 95 shows the temperature dependence of the diffusivity of lanthanum, and Fig. 96 its the thermal conductivity [140]. The uncertainty in the average values is about 10%. The electronic component is the main contributor to the

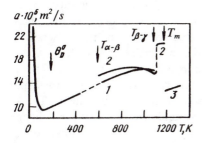

Figure 95 The diffusivity of lanthanum (a) vs. temperature: *1*) [39]; *2*) [140]; *3*) [155].

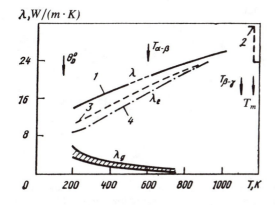

Figure 96 The thermal conductivity of lanthanum (λ) vs. temperature: *1*) [140]; *2*) [155]; *3*) $\lambda_e^g = \lambda - \lambda_g$; *4*) $\lambda_e^L = L_0T/\rho$; λ_g is the calculation of [140].

thermal conductivity [140]. At moderate temperatures, the values of λ_e^L and λ_e^g can be different because of the inelastic effects; but, at high temperatures, $\lambda_e^L = \lambda_e^g$.

The temperature dependence of the thermoelectric power (Fig. 97) reaches a maximum between 400 and 600 K. At higher temperatures, the thermoelectric power decreases and becomes negative [55]. Study [151] stated that the temperature dependences of the kinetic properties of scandium, yttrium and lanthanum have much in common, especially at high temperatures. The Hall coefficient of lanthanum changes from $-0.92 \cdot 10^{-10}$ m³/C at 170 K to $-0.3 \cdot 10^{-10}$ m³/C at 300 K [82].

§2 CERIUM, PRASEODYMIUM, NEODYMIUM, PROMETHIUM, SAMARIUM, EUROPIUM

CERIUM

At standard pressure, cerium has several polymorphic modifications. The low-temperature phase, α-Ce, has an hcp structure with the parameter $a = 0.485$ nm at 77 K. The data on the β-phase are contradictory. The α-β transformation takes place at 143 K, and may have a hysteresis because the β-phase can be supercooled to low temperatures. In addition, a magnetic transition is observed in this phase at 12.8 K. At 298 K, β-Ce has a double hcp structure with the parameters $a = 0.3673$ nm, $c = 1.1802$ nm. At 348 K, the β-phase transforms into the γ-phase, with an hcp structure. At 296 K, it has the parameter $a = 0.51601$ nm. The γ-phase can be easily supercooled, and at 160 ± 30 K it can be transformed into the α-phase. Below 13 K, the γ-Ce phase becomes magnetically ordered. The γ-α transformation is of an electronic type with a variable valence [156]; no change in the symmetry of the crystalline lattice takes place. At 983 K the γ-phase transforms into the δ-phase, which has a bcc structure with the parameter $a = 0.412$ nm at 1030 K. It should be noted that the phase transformation temperatures given here and in Table 6 are highly dependent on the purity and thermal history of the samples.

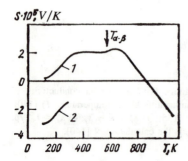

Figure 97 The thermoelectric power of lanthanum (S) vs. temperature: *1*) [151]; 2) data of Born, et al. [151].

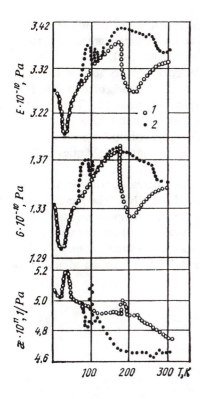

Figure 98 Young's modulus (E), shear modulus (G) and volume compressibility ($\varkappa$) of cerium vs. temperature [157]: *1*) heating; *2*) cooling.

The electronic spectrum of cerium has not been sufficiently investigated. An important feature of the spectrum is that the $4f$-level is probably located near the Fermi level for $6s$-$5d$-electrons [156, 41]. The Fermi surfaces of the high temperature phases γ-Ce and δ-Ce are not known.

The physical properties of cerium depend strongly on its thermomechanical history and its purity. The apparent temperature ranges of the three phases (α, β, γ) overlap. This makes the interpretation of the elastic characteristics of cerium difficult (Fig. 98). The investigated sample of 99.9% purity contained $\approx$50% of β-phase, and had a density of $(6.758 \pm 0.004)10^3$ kg/m^3 at T = 300 K.

On cooling down to $T_{\gamma-\beta} = 263$ K, the values of E and G increase sharply; then, the slopes of the $E(T)$ and $G(T)$ curves decrease. The curve of $\varkappa(T)$ has a peak near 170 K. The curves of $G(T)$ and $E(T)$ have wide maxima. A softening of the lattice takes place before the transition at $T \approx 110$ K. Two anomalies at 100 and 102 K are observed in the vicinity of the transition point. The reasons for their appearance are not clear. The 100 K phase transition is either related to the γ-α transition or is caused by the partial β-α transition. The nature of the minima of E and G at $T = 31$ K is also unclear. At $T_N = 13$ K the curves have only an inflection point. A large thermal hysteresis is typical for these dependences.

At 293 K, the following values were obtained for a cerium sample of 99.93% purity (γ-phase): $G = 1.235 \cdot 10^{10}$ Pa, $K = 1.885 \cdot 10^{10}$ Pa [158].

The temperature dependence of the electrical resistivity of cerium (Fig. 99) has a considerable nonlinearity with 'saturation' at high temperatures and hysteresis in the region of the phase transformations [140, 159, 160].

Curve 2 in Fig. 99 combines the data of different authors for the samples containing ≈0.1% impurities and a relative residual resistance of $r ≈ 20$. The uncertainty in the determination of ρ for γ- and β-phases is about 5%. The analysis of the electrical resistivity shows that the regular phonon contribution, calculated as for lanthanum, does not exceed 12% at 300 K. The principal contribution to the resistivity comes from the scattering by the magnetic inhomogeneities and by the localized 4f-levels.

Figure 100 shows that the diffusivity of cerium increases with temperature [39, 140]. The uncertainty in these results is ≈5% [140, 161]. The thermal conductivity is mainly of an electronic nature with a positive temperature coefficient [140]. The lattice component, calculated as for lanthanum, may amount to 30–50% of the total value at 200–300 K. However, its share decreases rapidly as the temperature increases. At high temperatures, the difference between λ_e^g and λ_e^L is

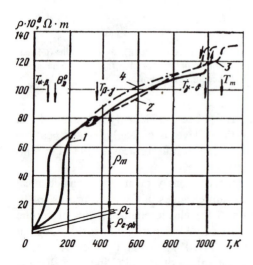

Figure 99 The electrical resistivity of cerium (ρ) vs. temperature: 1) [159]; 2) [140]; 3) [160]; 4) [151].

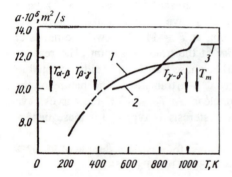

Figure 100 The diffusivity of cerium (a) vs. temperature: 1) [39]; 2) [140]; 3) [161].

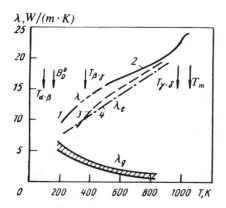

Figure 101 The thermal conductivity of cerium (λ) vs. temperature: *1*) [48]; *2*) [140]; *3*) $\lambda_e^g = \lambda - \lambda_g$; *4*) $\lambda_e^L = L_0 T/\rho$; λ_g is according to [140].

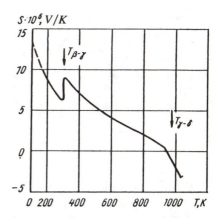

Figure 102 The thermoelectric power of cerium (*S*) vs. temperature [151].

small, which shows that the elastic scattering is dominant. At moderate temperatures, however, inelastic effects could also take place (Fig. 101).

The temperature dependence of the thermoelectric power is characterized by a negative slope and the presence of an inversion point [151, 160] (Fig. 102). At room temperature, the Hall coefficient of cerium is $R = 1.92 \cdot 10^{-10}$ m^3/C, and it decreases to $R = 1.10^{-10}$ m^3/C at 170 K [82].

The kinetic properties of cerium are summarized below [140, 151, 159–166].

T, K	200	300	400	500	600	700	800	900
$\rho \cdot 10^8$, $\Omega \cdot$ m	67	77	82	91	98	103	108	112
$a \cdot 10^6$, m^2/s	7.1	8.6	9.7	10.5	11:0	11.4	11.8*	12.3*
λ, W/(m$\cdot$K)	—	11.7	13.7	15.5	16.5	17.6	19.0	20.5

*The data require further investigation.

PRASEODYMIUM

At standard pressure, praseodymium has two allotropic modifications. The low-temperature phase, α-Pr, has a stable hcp structure with the lattice parameters $a = 0.36725$ nm, $c = 1.18354$ nm at 293 K. At 1069 K, this structure transforms into a bcc structure with $a = 0.413$ nm at 1094 K. At temperatures below 24 K, praseodymium transforms into a magnetically ordered state with a sinusoidal antiferromagnetic structure. The electronic structure and Fermi surface of praseodymium in the paramagnetic state are similar to those of lanthanum [41]. The elastic constants of praseodymium behave as expected during the decrease in temperature from 300 to 120 K (Fig. 103). The abrupt decrease in moduli below 80 K indicates that the lattice becomes softer; this precedes the antiferromagnetic ordering at 25

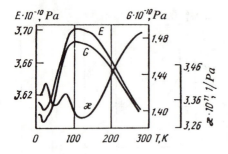

Figure 103 Young's modulus (E), shear modulus (G) and volume compressibility ($\varkappa$) of polycrystalline praseodymium vs. temperature [162].

K [162]. The elastic constants of praseodymium also become much softer below 120 K.

The elastic constants of praseodymium ($c_{ij} \cdot 10^{-10}$, Pa) are listed below [163].

T, K . .	10	20	30	80	120
c_{11} . . .	5.080	4.990	4.960	5.090	5.115
c_{33} . . .	5.570	5.490	5.420	5.645	5.755
c_{44}^a . . .	1.485	1.475	1.460	1.445	1.445
c_{44}^c . . .	1.520	1.510	1.490	1.460	1.450
c_{66} . . .	1.310	1.270	1.280	1.410	1.430
c_{12} . .	2.460	2.450	2.400	2.270	2.255
c_{13} . . .	1.48	1.54	1.58	1.52	1.47

T, K . .	160	200	240	280	300
c_{11} . . .	5.095	5.060	5.015	4.960	4.935
c_{33} . . .	5.795	5.795	5.780	5.750	5.740
c_{44}^a . . .	1.430	1.415	1.395	1.370	1.360
c_{44}^c . . .	1.435	1.420	1.400	1.371	1.358
c_{66} . . .	1.420	1.400	1.370	1.340	1.320
c_{12} . . .	2.255	2.260	2.275	2.280	2.295
c_{13} . . .	1.45	1.44	1.43	1.43	1.43

Note. The difference between the values of c_{14}, obtained from the measurements of velocity along the a- and c-axes, was caused either by the absence of the correction for thermal expansion of specimens in calculation of c_{ij}, or by the polarizing dispersion of c_{44}.

The magnetic ordering transition influences the temperature dependence of the electrical resistivity of praseodymium (Fig. 104). Figure 104 shows the data from various sources for polycrystalline praseodymium containing 0.3–0.1% of impurities.

The data need revision because the influence of the anisotropy in the crystalline structure on the resistivity is not known [140]. Nevertheless, all of the contributions to the electrical resistivity, typical for the transition metals, can be perceived in this case as well. The contribution from impurities is shown in Fig. 104 as $\rho_i = \rho_{4.2K}$. The phonon contribution was determined from the average slope of $\rho(T)$ in the high-temperature region. It is clear that, above the magnetic dis-

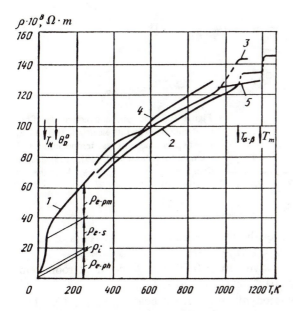

Figure 104 The electrical resistivity of praseodymium (ρ) vs. temperature: *1*) [164]; *2*) [165]; *3*) [151]; *4*) [140]; *5*) [166].

ordering point, the dependence of $\rho_m(T)$ is definitely nonlinear. Studies [140, 167] pointed to an anomaly in $\partial\rho/\partial T$ near 600 K, possibly due to an electron-electron phase transition. The diffusivity of polycrystalline praseodymium was studied for the both solid [140, 167] and liquid [166] states. Preliminary estimations of $a(T)$ [39] agree reasonably well with the results of [140]. The presence of an anomaly in the dependence of $a(T)$ (Fig. 105) [140, 167], and the difference in the locations of the minimum of $a(T)$, however, show that the obtained data are of a preliminary nature.

The information on the thermal conductivity of praseodymium was compiled in [140] and is shown in Fig. 106. When the lattice component of the thermal conductivity was calculated, the phonon-electron interaction was taken into account. The accuracy of the experiment did not permit the distinguishing of λ_e^g and λ_e^L. Generally speaking, however, the temperature dependence of λ is determined by its electronic component.

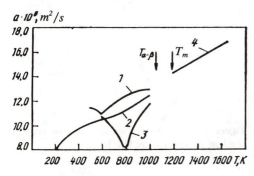

Figure 105 The diffusivity of praseodymium (*a*) vs. temperature: *1*) [140]; *2*) [39]; *3*) [167]; *4*) [166].

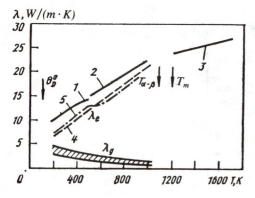

$\lambda, W/(m \cdot K)$

Figure 106 The thermal conductivity of praseodymium (λ) and its electronic (λ_e) and lattice (λ_g) components vs. temperature: *1–3*) the data for λ from [48], [39], [155], respectively; *4*) $\lambda_e^g = \lambda - \lambda_g$; *5*) λ_e^L; λ_g is the calculations of [140].

The temperature dependence of the thermoelectric power [151] is of an extreme nature, and above 700 K $S < 0$ (Fig. 107).

The Hall coefficient of praseodymium is $0.709 \cdot 10^{-10}$ m^3/C at room temperature, and decreases to $0.62 \cdot 10^{-10}$ m^3/C at 120 K [82].

The kinetic properties of praseodymium are listed below [39, 48, 140, 151, 154, 164].

T, K . . .	100	200	300	400	500	600	700	800	900
$\rho \cdot 10^8$, $\Omega \cdot$m	41	55	69	78*	89*	97*	104	111	117

T, K . . .	200	300	400	500	600	700
λ, W/(m$\cdot$K)	10.0	11.7	13.2	14.5*	16.2*	18.0*

T, K . . .	800	900	1000	1300	1400	1600
λ, W/(m$\cdot$K)	19.3	20.7	22.0	24.4	25.0*	26.3*

*The data need further investigation.

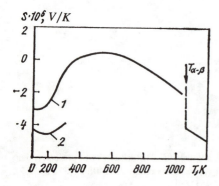

$S \cdot 10^6$, V/K

Figure 107 The thermoelectric power of praseodymium vs. temperature [151].

NEODYMIUM

At atmospheric pressure, neodymium has two polymorphic modifications [74]. The low-temperature α-phase has a double hcp structure with the lattice parameters being $a = 0.36579$ nm and $c = 1.17992$ nm at 293 K. The high-temperature β-phase has a bcc structure with $a = 0.413$ nm at 1156 K. The temperature of the α-β transformation is 1128 K. Below 19.55 K, neodymium is magnetically ordered and has a magnetic structure similar to that of praseodymium. Below 10 K, the temperature dependences of some properties of pure neodymium have anomalies, resulting from a change in the type of the magnetic ordering. This was not observed for impure samples.

In the paramagnetic region, the electronic structure and Fermi surface of neodymium should be similar to those of praseodymium and lanthanum. This, however, has to be verified [41].

Figure 108 shows the temperature dependences of the elastic constants of neodymium. The anomalies, such as the minima at 7.5 and 19 K, are caused by the magnetic reordering at these temperatures [168, 169].

In the paramagnetic phase, the behaviour of c_{ij} is regular with the exception that c_{66} has a stronger temperature dependence than c_{44}. Figure 109 shows the temperature dependences of Young's and shear moduli, and the volume compressibility for polycrystalline neodymium [162].

The temperature dependence of the electrical resistivity (ρ) of polycrystalline neodymium is similar to that of praseodymium. Most of the above results apply to polycrystalline neodymium containing 0.1–0.3% of impurities and a relative residual resistance of $r \approx 10$ [140, 170–172]. The resistivity of single crystal neodymium is not known. The data shown in Fig. 110 may reflect the influences of the anisotropy and the thermomechanical history of the samples. Therefore, further investigations are needed.

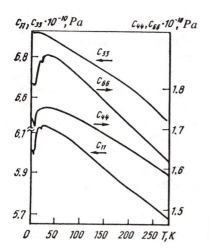

Figure 108 The elastic constants of neodymium vs. temperature [163].

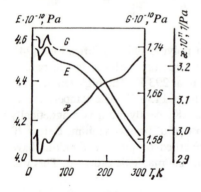

Figure 109 Young modulus (E), shear modulus (G) and volume compressibility ($\varkappa$) of polycrystalline neodymium vs. temperature [162].

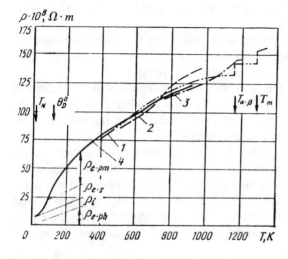

Figure 110 The electrical resistivity of neodymium (ρ) vs. temperature: *1*) [165]; 2) [167]; 3) [151]; 4) the average data of [140, 151, 164–167, 170–172].

The components of the electrical resistivity of neodymium were calculated in the same manner as for praseodymium. It was shown that $\rho_{e-pm} \approx \rho_{e-s}$, although the value of ρ_{e-s} in neodymium is somewhat smaller than in praseodymium. Data of [140] and [167] indicate an anomaly in the electrical resistivity near 700 K. This anomaly is strongly influenced by the impurities present in a sample. The anomaly is more sharply pronounced in the temperature dependence of diffusivity (Fig. 111). Studies [140, 167] pointed out that these anomalies are related neither to a substantial change in the symmetry of the crystalline structure nor to the magnetic disordering. Additional experiments are needed to determine the nature of these anomalies.

At 300 K, the contribution of the lattice thermal conductivity to the overall thermal conductivity of neodymium (Fig. 112) amounts to about 30%. It decreases rapidly, however, as the temperature increases. It is difficult to notice a difference between λ_e^L and λ_e^g because the limits of uncertainty are 10–15%.

The temperature dependence of the thermoelectric power of neodymium reaches

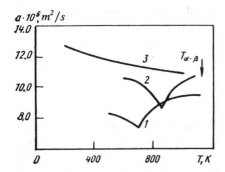

Figure 111 The diffusivity of neodymium (*a*) vs. temperature: *1*) [140]; *2*) [167]; *3*) [39].

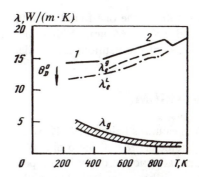

Figure 112 The thermal conductivity of neodymium (λ) vs. temperature: *1*) [48]; *2*) [140].

a maximum at 600–700 K (Fig. 113). In [151] a jump in the value of $S(T)$ was registered during the α-β transformation.

At 293 K, the Hall coefficient of neodymium is equal to $0.97 \cdot 10^{-10}$ m³/C. As temperature decreases, the Hall coefficient increases monotonically, reaching $1.3 \cdot 10^{-10}$ m³/C at 40 K [82, 168].

The kinetic properties of neodymium are listed below [39, 140].

T, K ...	100	200	300	400	500	600	700	800
$\rho \cdot 10^8$, $\Omega \cdot$ m	26	50	68	80	90	98	107	115

T, K	300	400	500	600	700	800
λ, W/(m·K) ·	14.4	14.5	15.2	15.8	16.5	17.2

PROMETHIUM

Promethium does not have any stable isotopes (the half-life of the most long-lived isotope ^{145}Pm is equal to 30 years). Its physical properties are very poorly known. Study [82] indicates that, at room temperature, promethium has a hexagonal struc-

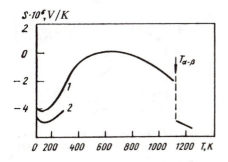

Figure 113 The thermoelectric power of neodymium (S) vs. temperature: *1*) [151]; *2*) the data of Born, et al. [151].

ture similar to the one found in lanthanum, with the lattice parameters $a = 0.365$ nm and $c = 1.165$ nm. With some caution it can be assumed that the physical properties of promethium are similar to those of neodymium and samarium.

SAMARIUM

At temperatures below 1197 K, samarium [82] has a rhombohedral structure with the parameters $a = 0.8996$ nm, $\beta = 23.21°$ (according to [74], samarium has a hexagonal structure of its own type with $a = 0.3621$ nm, $c = 2.635$ nm); above 1197 K, it has a bcc structure with $a = 0.407$ nm. Below 105 K, samarium is magnetically ordered; from 105 to 13.3 K it is antiferromagnetic; and, below 13.3 K, it is ferromagnetic.

The electronic structure of samarium has not been sufficiently studied. The temperature dependences of the Young's and shear moduli, and of the volume compressibility of polycrystalline samarium are shown in Fig. 114. The magnetic ordering at $T_N = 105$ K is distinguished by a peak in the curve of $\varkappa(T)$ and inflection points in the curves of $E(T)$ and $G(T)$. At 14 K, during the transition into the ferromagnetic state, the crystal lattice softens [162].

The information on the electrical resistivity of polycrystalline samarium is summarized in Fig. 115 and listed below [140, 167].

T, K . . .	100	200	300	400	500	600	700	800	900
$\rho \cdot 10^8$, $\Omega \cdot$ m	64*	77*	90	100	111	123*	140*	153*	164*

*The data need verification.

The existence of a linear region, above the Neel point, should be mentioned. This indicates that a considerable contribution to the electrical resistivity comes

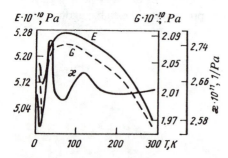

Figure 114 The elastic modulus (E), shear modulus (G) and volume compressibility ($\varkappa$) of polycrystalline samarium vs. temperature [162].

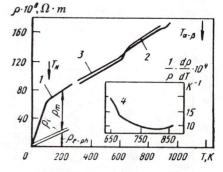

Figure 115 The electrical resistivity of samarium (ρ) vs. temperature: *1*) [170]; *2*) [151]; *3*) [140]; *4*) [174].

from the scattering by disordered spins. At 600–700 K the temperature dependence of the resistivity has an inflection point, and the values of $\partial\rho/\partial T$ reach a maximum [140, 167]. This anomaly was investigated in samples of different purity [140, 173, 174].

Generally speaking, the data for the electrical resistivity of samarium need revision because the influence of the anisotropy of crystalline structure is not known. In addition, the temperatures of the phase transitions depend appreciably on the sample purity (for example, studies [151, 173] mentioned two anomalous points). The diffusivity data for samarium [140] also indicate a high-temperature anomaly above 600 K.

Figure 116 shows the results of diffusivity calculations for low temperatures [39]. The anomaly near the Neel point appears as a distinct minimum. The data on the diffusivity of samarium are given below.

T, K	300	400	500	600
$a \cdot 10^6$, m²/s	9.2	8.5	7.7	7.1

Figure 117 shows the thermal conductivity of samarium; the general shape of the temperature dependence is well established. Above 300 K, the values of λ increase with temperature. The positive temperature coefficient of the thermal conductivity is due to its electronic component, even though the values of λ_e^g and λ_e^L are appreciably different for samarium. As a result its electronic Lorenz number, at high temperatures, is greater than the standard value of L_0 because of the inelastic contributions. The values of the thermal conductivity at different temperatures are listed below [140, 167].

T, K	100	200	300	400	500	600	700	800	900
λ, W/(m·K)	9.0	11.0	12.8	14.1	15.2	16.1	17.0	17.6	18.3

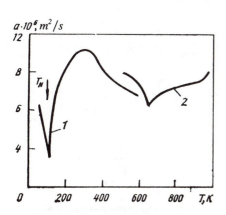

Figure 116 The diffusivity of samarium (a) vs. temperature: *1*) [39]; *2*) [140].

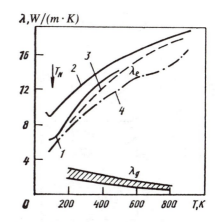

Figure 117 The thermal conductivity of samarium (λ) vs. temperature: *1*) [173]; *2*) the average of [48, 173, 176]; *3*) $\lambda_e^g = \lambda - \lambda_g$; *4*) $\lambda_e^L = L_0 T/\rho$; λ_g is according to [140].

The temperature dependence of the thermoelectric power of polycrystalline samarium has a shape that is typical for the light rare earth metals: a maximum at moderate temperatures and a decrease into the region of negative values at high temperatures (Fig. 118). Study [151] mentioned the presence of an anomaly in $S(T)$ at 600–700 K.

The Hall coefficient of samarium is positive below 106 K; reaches a maximum value of $2.8 \cdot 10^{-10}$ m^3/C at 20 K; and, has an inversion point near 110 K. Above 300 K, the Hall coefficient is almost constant and equal to $-0.4 \cdot 10^{-10}$ m^3/C up to 700 K. In the region of 750–770 K a small jump in the value of the Hall coefficient is observed (from $-0.38 \cdot 10^{-10}$ to $-0.5 \cdot 10^{-10}$ m^3/C) [168, 199].

EUROPIUM

The crystalline structure of europium differs sharply from that of other light rare earth metals. Up to the melting point it has a bcc structure with the parameter $a = 0.4582$ nm at 298 K. Below 90 K, europium transforms into the antiferromagnetic state with a helical antiferromagnetic structure, the axes of which are parallel to the edges of a cube.

The Fermi surface of europium is similar to that of barium. The moduli of elasticity of polycrystalline europium (samples of 99.9% purity) have been measured for the temperature interval from 4 to 300 K [177]. Europium differs from the other rare earth metals by its high volume compressibility; at $T = 300$ K $\varkappa = 1.2 \cdot 10^{-10}$ 1/Pa. The magnetic ordering at $T_N = 90$ K appears as a λ-type anomaly in the temperature dependences of G, E and $\varkappa$ (Fig. 119). The wide maximum in the curve of $\varkappa(T)$ near 150 K correlates with the distinctive features

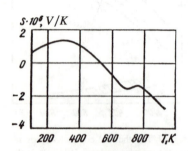

Figure 118 The absolute thermoelectric power of samarium (S) vs. temperature [151].

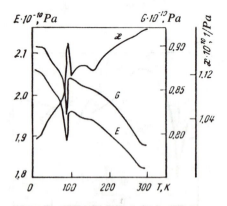

Figure 119 Young's modulus (E), shear modulus (G) and volume compressibility (x) of polycrystalline europium vs. temperature [177].

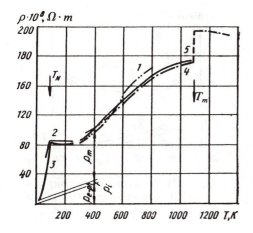

Figure 120 The electrical resistivity of europium (ρ) vs. temperature: *1*) [140]; *2*) [164]; *3*) [170]; *4*) [179]; *5*) the average of [140, 164, 179–181].

in the curves of $\rho(T)$ and $1/\varkappa(T)$. The author of [177] proposed the following explanation: the unusual elastic parameters near 150 K are caused by a change in the electronic structure of europium resulting from inter-band electronic transitions.

The temperature dependence of the electrical resistivity of polycrystalline europium is given in Fig. 120 and below.

T, K	100	200	300	400	500	600	700	800	900	1000
$\rho \cdot 10^8$, $\Omega \cdot$ m	84	81	86	102	120	140	159	175*	186*	193*

*The data require further investigation.

Europium does not have anisotropy in resistivity because of its cubic structure. Therefore, most of the literature data [140, 179, 180] are similar (the difference does not exceed 10%). The important features are a maximum at the Neel temperature and an inflection point corresponding to a maximum in $\partial\rho/\partial T$ above 400 K.

The regular phonon contribution to the electrical resistivity was isolated on the basis of the resistivity of barium as both elements have similar crystalline and electronic structures. The magnetic component of resistivity has a wide minimum at 300 K. It then increases with temperature, reaching 'saturation' at 1000 K. The anomaly in the temperature dependence of the electrical resistivity is caused by the anomaly in its magnetic component.

According to [140], the diffusivity of europium has a minimum near 600 K (Fig. 121).

The total thermal conductivity of europium is approximately equal to that of its electronic component, which determines the positive temperature coefficient. The uncertainty in the values of λ, given below and in Fig. 122, is 10–15%.

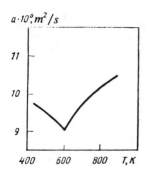

Figure 121 The diffusivity of europium (a) vs. temperature [140].

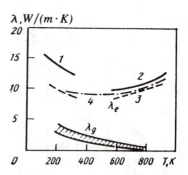

Figure 122 The thermal conductivity of europium (λ) vs. temperature: *1*) [48]; *2*) [140]; *3*) λ_e^g; *4*) λ_e^L; λ_g is according to [140].

T, K	500	600	700	800	900
λ, W/(m·K) . .	9.5	10.0	10.5	11.5*	13.0*

*The data require further investigation.

The thermoelectric power of europium differs substantially from that of the other rare earth metals. According to the data of [151, 179], this dependence has a wide maximum in the interval of 500–700 K. However, even in the liquid state, the thermoelectric power values remain positive (Fig. 123). At room temperature, the Hall coefficient of europium is equal to $25 \cdot 10^{-10}$ m^3/C [168].

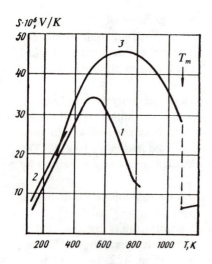

Figure 123 The absolute thermoelectric power of europium (S) vs. temperature: *1*) [151]; *2*) the data of Meaden and Sze referred to in [151]; *3*) [179].

§3 GADOLINIUM, TERBIUM, DYSPROSIUM, HOLMIUM, ERBIUM, THULIUM, YTTERBIUM, LUTETIUM

GADOLINIUM

At standard pressure, gadolinium has two polymorphic modifications: an hcp modification below 1535 K with the lattice parameters $a = 0.36360$ nm, $c = 0.57826$ at 293 K, and a bcc modification above 1535 K with $a = 0.406$ nm [74]. Gadolinium is ferromagnetic below 291.8 K. Another magnetic phase transition, caused by a change in the spin orientation, takes place at 228 K.

The results of calculating the Fermi surface and electronic spectrum of gadolinium are shown in Figs. 124 and 125. Fig. 124 shows that the Fermi level lies near the maximum of the density of states (to the left of the maximum, hence $\partial N/\partial \varepsilon > 0$). The Fermi surface is rather complex (Fig. 125), and it is mainly in the third and fourth zones of the double-zone scheme.

Figure 126 shows the elastic constants of single crystal gadolinium ($r = 51$) [182]. The magnetic ordering and restructuring manifest themselves as an anomalous behaviour of the constant c_{33} (a sharp minima in the vicinity of $T_c = 292$ K and $T_s = 228$ K). At $T \approx 180$ K, when the angle of rotation of magnetic moments with respect to the c-axis reaches a maximum, the curves of c_{11} and c_{33} exhibit cusp points.

The elastic constants of gadolinium for temperatures from 0 to 360 K are listed below [183]. The quoted values of $c_{ij} \cdot 10^{-10}$ Pa differ from those given in [182] because the samples had different purity and, probably, different orientations.

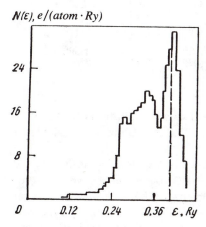

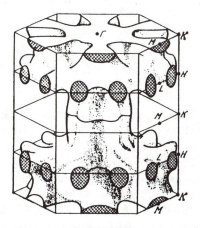

Figure 124 The density of the electronic states of gadolinium [$N(\varepsilon)$] as a function of energy [181].

Figure 125 The Fermi surface of gadolinium according to the double-zone scheme (the cross sections of holes are shaded) [41].

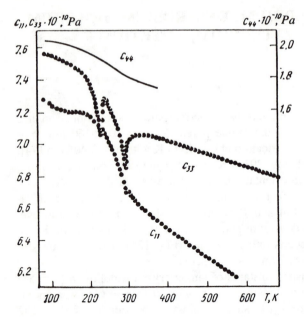

Figure 126 The elastic constants of gadolinium (c_{ij}) vs. temperature [182].

T, K . .	0	40	80	120	160	200	240
c_{11} . . .	7.680	7.653	7.582	7.525	7.487	7.398	7.233
c_{33} . . .	7.899	7.856	7.795	7.725	7.667	7.540	7.372
c_{44} . . .	2.378	2.353	2.345	2.318	2.288	2.249	2.193
c_{12} . . .	—	—	—	—	—	—	2.745
c_{13} . . .	1.910	1.917	1.928	1.942	1.933	1.937	1.957

T, K . .	280	290	295	300	320	360
c_{11} . . .	6.962	6.833	6.800	6.783	6.722	6.612
c_{33} . . .	6.990	6.820	7.050	7.123	7.148	7.140
c_{44} . . .	2.113	2.093	2.084	2.077	2.058	2.037
c_{12} . . .	2.631	2.593	2.576	2.559	2.516	2.502
c_{13} . . .	1.976	2.100	2.086	2.072	2.065	2.080

The temperature dependences of Young's and shear moduli for polycrystalline gadolinium of 99.9% purity are given in Fig. 127. The values were calculated with a correction for the thermal expansion of samples.

Figure 128 shows the temperature dependence of the electrical resistivity for single crystal and polycrystalline gadolinium, based on the data given in compilation [140]. Below 300 K the data are taken from [185]; and, in the interval 300 to 1100 K, from [140]. Above 1100 K, the data of [149] for polycrystalline sample were used because the difference between $\rho_\perp$ and $\rho_\parallel$ practically disappears. The resistivity of the polycrystalline sample was calculated as $\rho_{poly} = 2/3\rho_\perp + 1/3\rho_\parallel$. Figure 128 and Table 22 give the data for 99.99% pure gadolinium with a relative residual resistance of $r \approx 40$ [140]; the uncertainty of determination

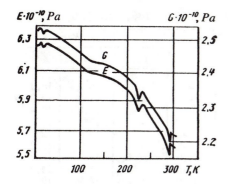

Figure 127 Young's (E) and shear (G) moduli of polycrystalline gadolinium vs. temperature [184].

is $\approx 3\%$. Figure 129 shows the derivative of resistivity with respect to temperature [189]. Sharp anomalies are observed in the vicinity of the magnetic transitions in the direction of the a-axis. In the direction of the c-axis, the anomaly near T_s is less pronounced. Figure 128 also shows the temperature dependence of the anisotropy $\rho_\perp/\rho_\parallel$ which has a maximum near 500 K and disappears at about 1100–1200 K.

The phonon contribution to resistivity was calculated by averaging the slope of $\rho(T)$ over the interval of 400–700 K. Figure 129 shows that the magnetic portion amounts to about 80% of the total resistivity at 300 K. Kasuya [186]

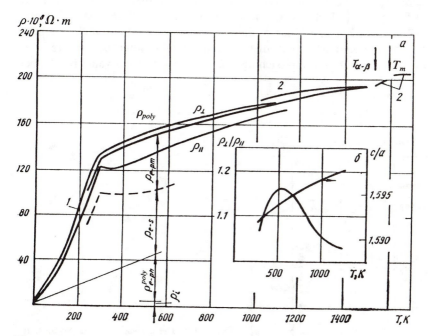

Figure 128 The temperature dependences of a) electrical resistivity of gadolinium (ρ), b) anisotropy in resistivity ($\rho_\perp/\rho_\parallel$) and in the value of c/a [140]: 1) [185, 140]; 2) [149].

Table 22 The kinetic properties of gadolinium [140, 149, 188, 189]

T, K	$\rho \cdot 10^{8}, \Omega \cdot m$			$a \cdot 10^{9}, m^{2}/s$			$\lambda, W/(m \cdot K)$	
	poly	‖ c	⊥ c	poly	‖ c	⊥ c	‖ c	⊥ c
100	—	—	—	—	—	—	14.2	15.0
200	—	—	—	—	—	—	11.8	12.2
300	130.3	127.0	136.2	—	—	—	11.0	10.4
400	139.0	126.2	147.1	—	—	—	13*	11*
500	145.6	132.3	154.3	—	—	—	14*	12*
600	152.0	139.4	160.5	—	—	—	14.9	12.3
700	157.0	146.6	166.0	—	10.20	8.60	15.6	13.3
800	162.8	154.6	171.1	—	10.45	9.01	16.4	14.3
900	167.7	163.0	175.2	9.7	10.70	9.45	16.9	15.3
1000	172.0	169.1	178.8	10.0	10.95	9.85	17.8	16.3
1200	178.3	176.0	184.0	10.6	11.20	10.60	20.0	18.4
1400	—	—	—	11.2	11.15	11.10	21.0	20.3
1500	—	—	—	11.0*	11.0*	11.3*	21.5*	21.0*

*The data require further investigation.

calculated the magnetic contribution from scattering by the disordered f-spins ρ_{e-s} (see Fig. 128). The residual portion of resistivity $\Delta_{\rho} = \rho - \rho_{e-s} - \rho_{e-ph} - \rho_{l}$ is probably due to other types of scattering (scattering of electrons by parmagnons and localized spin fluctuations). Study [140] pointed out that at high temperatures the mean free path of electrons becomes comparable to the interatomic distance. This may cause the disappearance of the anisotropy in resistivity and a decrease in $\partial \rho / \partial T$. It is important to note that the ratio of the lattice parameters, c/a, increases with temperature, approaching $c/a = 1.63$ (the ideal value for the hcp lattice).

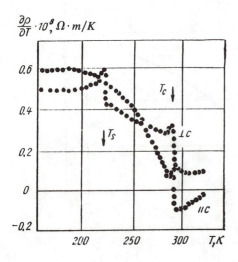

Figure 129 The derivative of the electrical resistivity of gadolinium with respect to temperature [189].

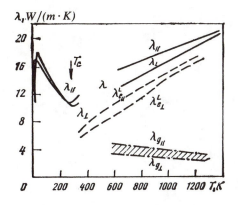

Figure 130 The thermal conductivity of gadolinium (λ) vs. temperature [140, 188]; within the error of estimation $\lambda_e^L = \lambda_e^g$

The diffusivity of single crystals of gadolinium is given in Table 22 [140]. For the solid state, within the limits of 10%, the values overlap with the polycrystal data of [155]. At 1200–1300 K the anisotropy in diffusivity disappears.

Figure 130 shows the thermal conductivity of single crystals of gadolinium ($r \approx 30$–50). The temperature dependence is of an extreme nature. A low-temperature maximum is usual for crystalline phases. Near the Curie point the thermal conductivity has a minimum, then its values increase with temperature.

Figure 131 shows the results of careful measurements of $\lambda(T)$ near T_c for polycrystalline gadolinium ($r = 86$). The minimum due to the critical scattering by phonons is visible despite some increase in the thermal conductivity in the interval $T_c + 1$ K. Calculations of the lattice component of the thermal conductivity show that, at low and moderate temperatures, it may reach 50% of the total value. The electronic component, calculated at $\lambda_e^g = \lambda - \lambda_g$, indicates the presence of appreciable inelastic contributions because $\lambda - \lambda_g > L_0 T/\rho$; hence, the Lorenz number exceeds the standard value of L_0 (Fig. 132). When the temperature

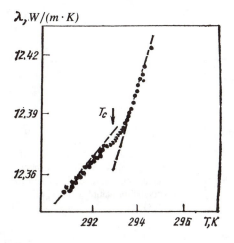

Figure 131 The thermal conductivity of polycrystalline gadolinium vs. temperature near the Curie point [187].

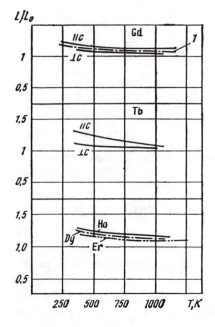

Figure 132 The Lorenz function of metals of the gadolinium subgroup vs. temperature [140].

increases, $L \rightarrow L_0$, and above 1000 K $L \approx L_0$. At about the same temperature, the anisotropy in the thermal conductivity disappears.

The thermoelectric power of gadolinium has a complex nonmonotonical temperature dependence with small anomalies near the T_s and T_c points (Fig. 133). The important features of this dependence are a maximum at 700–800 K and a decrease in the thermoelectric power at higher temperatures, while a substantial anisotropy is retained.

The Hall coefficient of gadolinium is negative and its absolute value decreases (Fig. 134).

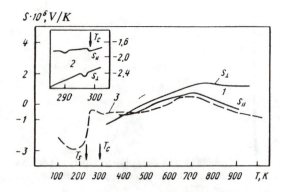

Figure 133 The absolute thermoelectric power of gadolinium (S) vs. temperature; for a single crystal (1, 2) and for a polycrystalline sample (3): 1) [153]; 2) [33]; 3) [151].

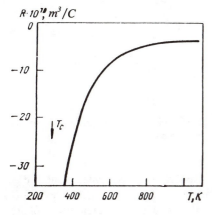

Figure 134 The Hall coefficient of polycrystalline gadolinium vs. temperature [151].

TERBIUM

At standard pressure, in the paramagnetic region, terbium has an hcp structure with the latice parameters $a = 0.36010$ nm and $c = 0.56936$ nm at 293 K. Above 1560 K, terbium has a bcc structure with $a = 0.402$ nm [39, 74]. Terbium has two magnetic transitions: the Neel point at 227.5 ± 1 K and the Curie point at 220 ± 2 K. Below 220 K, a rhombohedral modification of terbium can exist [74].

The electronic structure of terbium is similar to gadolinium (Fig. 125). According to the results of calculations in [41], the Fermi surface of the elements from gadolinium to other heavy rare earth metals changes. The changes include the disappearance of the two 'arms' at point M and the merging of the two 'arms' at the L point. For terbium, the 'arms' at the M point remain unchanged; however, they probably merge at the L point.

The behavior of the elastic parameters of terbium is very similar to that of dysprosium, except that terbium is a harder material. The elastic constants of the hexagonal terbium, $c_{ij} \cdot 10^{-10}$ Pa, are listed below [183].

T, K	0	40	80	120	160	200	215
c_{11}	—	—	—	6.618	5.800	5.888	6.268
c_{33}	8.243	8.252	8.170	8.061	7.878	7.492	7.245
c_{44}	2.518	2.511	2.488	2.447	2.373	2.298	2.270
c_{12}	—	—	—	—	—	—	—
c_{13}	—	—	—	—	—	—	—

T, K	220	225	230	240	280	300
c_{11}	6.610	7.173	7.130	7.115	6.973	6.924
c_{33}	7.216	7.150	7.193	7.351	7.435	7.439
c_{44}	2.261	2.251	2.242	2.225	2.189	2.175
c_{12}	2.870	2.880	2.606	2.475	2.483	2.498
c_{13}	—	2.094	—	2.175	—	2.179

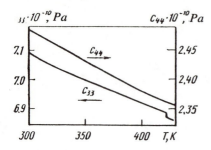

Figure 135 The elastic constants c_{33} and c_{44} of terbium vs. temperature [190].

The behavior of c_{11} indicates on structural distortions in the basal plane [183].

Figure 135 shows the temperature dependences of c_{33} and c_{44} above 300 K for terbium with $r \approx 30$. The behavior of c_{33} at $T = 420$ K is related to a phase transition of the order of 2.5 [190]. Figure 136 shows the elastic moduli and the volume compressibility for polycrystalline terbium of 99.9% purity for the temperature range from 4 to 300 K. The anomalies near the magnetic transition points have a somewhat unusual shape, the typical minima in $G(T)$ near T_N and in $E(T)$ near T_c are absent.

The electrical resistivity of polycrystalline and single crystal terbium is given in Figs. 137–139 and Table 23. Below 300 K, the data of Volkenshtein, et al, [14] have been used.

Studies [190, 192] showed that, near 420 K, peculiarities may appear in the temperature dependence of the electrical resistivity and other properties. The anisotropy in resistivity reaches 1.27 at 400 K, decreases monotonically with temperature, and disappears below 1200–1300 K. At the same time, the lattice parameter ratio approaches the ideal value ($c/a = 1.63$).

The above data for the electrical resistivity represent a metal with $r \approx 40$. The phonon contribution was estimated from the average slope in the resistivity over the interval 300–700 K. The residual portion of the resistivity, $\rho_m = \rho - \rho_i - \rho_{e-ph}$, is due to the electron-electron scattering. Kasuya [186] calculated the magnetic contribution to resistivity of paramagnetic terbium from the electron scattering by disordered spins, ρ_{e-s} (see Fig. 137). However, as in the case of

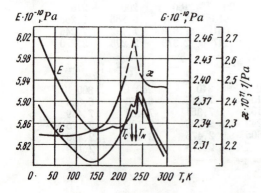

Figure 136 Young's modulus (E), shear modulus (G) and volume compressibility ($\varkappa$) of polycrystalline terbium vs. temperature [184].

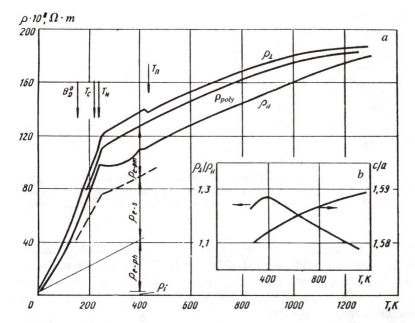

Figure 137 The temperature dependence of *a*) the electrical resistivity of terbium (ρ) [14, 140, 188], *b*) the anisotropy in the electrical resistivity ($\rho_\perp/\rho_\parallel$) and the lattice parameter ratio c/a.

gadolinium, a visible portion of resistivity, $\rho_m - \rho_{e-s}$, still exists. This portion can be explained by other scattering mechanisms involving the *f*-levels or paramagnons (the latter portion,is designated in Fig. 137 as ρ_{e-pm}). Thus, the magnetic portion of resistivity is responsible for the specific high temperature behavior of resistivity.

The information of diffusivity of terbium, given in Fig. 140 and Table 23, corresponds to the results of [140] for single crystal and polycrystalline samples

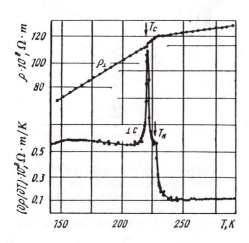

Figure 138 The anomalies in the electrical resistivity of terbium (ρ) and its derivative with respect to temperature in the vicinity of the Curie and Neel points, in the direction perpendicular to the hexagonal axis [191].

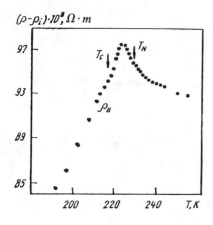

$(\rho - \rho_i) \cdot 10^9, \Omega \cdot m$

Figure 139 The anomaly in the ideal electrical resistivity of terbium, $\rho - \rho_i$, near the Curie and Neel points, in the direction parallel to the hexagonal axis [191].

with $r = 30$, and to calculations in [39] for a metal of similar purity. The uncertainty in these values is $\approx 5\%$, above 800 K, and 10–20% at low and moderate temperatures. Near the magnetic transition points, the observed minimum in the value of a corresponds to a maximum in the specific heat. Above 1300 K, the anisotropy in diffusivity disappears.

Figure 141 and Table 23 show the thermal conductivity of single crystal terbium. As before, the lattice component was calculated with the phonon-electron scattering taken into consideration. At low temperatures, the lattice component has the same order of magnitude as the electronic component, and the Lorenz number exceeds the value of L_0. The difference vanishes as the temperature increases (Fig. 142).

The thermoelectric power of single crystal terbium has distinctive features in

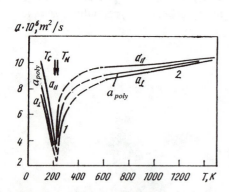

Figure 140 The temperature dependence of diffusivity (a) for single crystal terbium in the direction of the c-axis ($a_\parallel$), and perpendicular to it ($a_\perp$), and for a polycrystalline sample (a_{poly}) [39, 140]: *1*) [39]; *2*) [140].

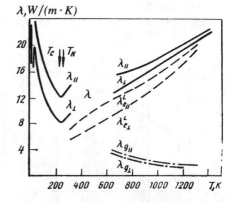

Figure 141 The thermal conductivity (λ) of single crystal terbium in the direction of the c-axis ($\lambda_\parallel$), perpendicular to it ($\lambda_\perp$) [140, 188], and for a polycrystalline sample (λ_{poly}) vs. temperature [140]; λ_g is calculated from [140].

Table 23 The kinetic properties of terbium [140, 188, 190–192]

T, K	$\rho \cdot 10^8$, $\Omega \cdot m$			$a \cdot 10^6$, m^2/s			λ, W/(m·K)	
	poly	‖c	⊥c	poly	‖c	⊥c	‖c	⊥c
100	—	30	40	—	—	—	—	—
200	87	76	94	—	—	—	—	—
300	117	98	127	7.3	8.2	6.4	13.5*	9.3*
400	128	110	138	8.3	9.0	7.6	14.0*	10.5*
500	137	118	146	8.7	9.4	8.2	14.8*	11.4*
600	147	128	154	9.0	9.6	8.6	15.0	13.0
700	156	137	162	9.15	9.7	8.8	15.6	13.9
800	164	146	170	9.3	9.8	9.0	16.4	14.9
900	170	154	176	9.4	9.8	9.2	17.1	15.7
1000	176	162	181	9.6	9.9	9.3	18.0	17.0
1200	184	177	188	9.8	10.0	9.6	19.1	18.9
1400	—	—	—	—	10.2	10.1	21.0	20.0

*The data obtained by the interpolation of results of [140] and [188].

the vicinity of the magnetic transition points (Fig. 143) [151]. At higher temperatures, it increases, reaching a maximum near 500–800 K. According to [153], the subsequent decrease in the thermoelectric power and a simultaneous disappearence of the anisotropy can be explained by the decrease in the mean free path of electrons, which approaches the interatomic distance.

The Hall coefficient of terbium is negative, indicating a predominantly electronic conductivity. Its absolute value decreases with increases in temperature (Fig. 144) [151].

DYSPROSIUM

At standard pressure, dysprosium has a tightly packed hcp structure with the lattice parameters $a = 0.35903$ nm and $c = 0.56475$ nm at 293 K. This modification

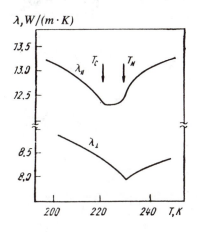

Figure 142 The anomaly in the thermal conductivity of terbium (λ) near the Curie and Neel points [188].

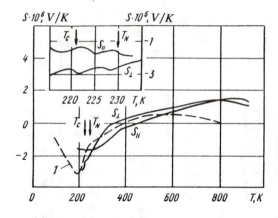

Figure 143 The absolute thermoelectric power (S) vs. temperature for single crystal terbium in the direction of the c-axis ($S_{\parallel}$), perpendicular to it ($S_{\perp}$) [153, 192], and for a polycrystalline sample [151] (broken line).

transforms at 1657 K into a bcc structure with $a = 0.398$ nm. The Neel point of dysprosium is 177.5 K, and the Currie point -83.5 K. A stable rhombic modification can exist in the ferromagnetic state [74].

The electronic structure of dysprosium resembles that of gadolinium. The Fermi energy lies near the maximum of the density of the electronic states $N(\varepsilon)$. The Fermi surface is intermediate between those of gadolinium and yttrium [41].

Figure 145 and Table 24 give the temperature dependence of the elastic constants of dysprosium. The magnetic ordering at $T_N = 179$ K causes a considerable softening of the lattice in the direction of the hexagonal axis [the minimum in $c_{33}(T)$]. The transition into the ferromagnetic phase at $T_c = 87$ K is accompanied by the 'hardening' of the lattice in the direction of the c-axis and softening in the basal plane. The rhombic distortion of the lattice at T_c gives rise to considerable changes ($\approx 20\%$) in the elastic moduli of polycrystalline dysprosium. Figure 146 shows the temperature dependence of the elastic characteristics of dysprosium of 99.9% purity.

Figure 147 shows the temperature dependence of the electrical resistivity of dysprosium. The low temperature data for single crystal dysprosium are taken from [194]. However, for polycrystalline dysprosium above 300 K the only available data are [140, 195]. Figure 147 shows that the results of different studies

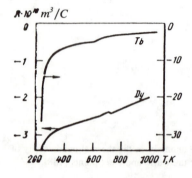

Figure 144 The Hall coefficient of terbium and dysprosium (R) vs. temperature [151].

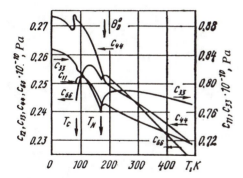

Figure 145 The elastic constants of dysprosium vs. temperature [38].

agree within 10%. Study [140] found an inflection point in $\rho(T)$ with a maximum of $\partial\rho/\partial T$ near 500–600 K. The results for the average dependence of $\rho(T)$ apply to a metal with $r \approx 30$.

The contributions to the electrical resistivity from different mechanisms of scattering are also shown in Fig. 147. The magnetic contribution was determined as $\rho_m = \rho - \rho_i - \rho_{e-ph}$ [196], where ρ_{e-ph} was found from the slope of the temperature dependence $\rho_{e-ph} = AT$ in the interval 300–800 K. At high temperatures, the value of ρ_m, obtained as described above, decreases.

In [186] the contribution from the electron scattering by disordered spins, ρ_{e-s}, was calculated. It amounts to about 50% of ρ_m. The remaining part of ρ_m could be explained in the same manner as it was for gadolinium and terbium.

The temperature dependence of the diffusivity of dysprosium is given in Fig. 148. At low temperatures, it corresponds to the results of calculation in [39], according to which the curve of diffusivity has minima at the magnetic phase transition points. The differences among the high-temperature data are 10%. The data require a correction for the anisotropy of the crystalline structure.

Table 24 The elastic constants of dysprosium ($c_{ij} \cdot 10^{-10}$, Pa)

T, K	c_{11}	c_{33}	c_{44}	c_{12}	c_{13}	T, K	c_{11}	c_{33}	c_{44}	c_{12}	c_{13}
According to the data of [193]						According to the data of [193]					
						250	7.42	7.83	2.43	2.55	2.25
						300	7.31	7.81	2.40	2.53	2.23
4.2	8.01	8.51	2.68	—	1.86	According to the data of [38]					
50	7.79	8.46	2.66	—	1.79						
80	—	8.36	2.60	—	—	298	7.466	7.871	2.427	2.616	2.233
90	—	8.18	2.57	—	—	373	7.337	7.803	2.378	2.641	2.218
100	7.81	8.14	2.67	2.91	1.75	473	7.161	7.710	2.298	2.669	2.202
140	7.85	7.97	2.61	2.81	1.99	573	6.984	7.598	2.222	2.708	2.184
160	7.73	7.84	2.57	2.72	1.85	673	6.810	7.488	2.14	2.740	2.168
170	7.65	7.74	2.53	2.67	2.19	773	—	7.363	2.076	—	—
180	7.56	7.70	2.49	2.60	2.35	873	—	7.223	2.006	—	—
200	7.52	7.80	2.47	2.56	2.30	923	—	7.149	1.970	—	—

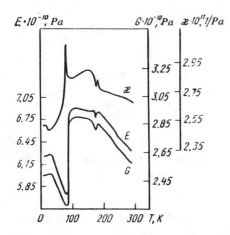

Figure 146 Young's modulus (E), shear modulus (G) and volume compressibility ($\varkappa$) of polycrystalline dysprosium vs. temperature [184].

At high temperatures, the increase in the thermal conductivity, shown in Fig. 149, is due to its electronic component, and, at moderate temperatures, it is due to the inelastic contributions. Some deviation of the Lorenz number from the standard value is possible (see Fig. 132).

According to [151], the thermoelectric power of dysprosium has a smooth maximum near 800–900 K and sharp anomalies at the magnetic phase transition points (Fig. 150). The Hall coefficient (see Fig. 144) is negative and its absolute

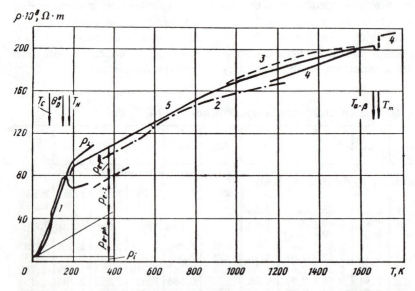

Figure 147 The electrical resistivity of dysprosium vs. temperature: *1*) [194], for a single crystal in the direction of the *c*-axis ($\rho_\parallel$), and perpendicular to it ($\rho_\perp$); *2, 3, 4*) the data of [140], [195], and [141], respectively, for polycrystalline dysprosium; *5*) the average data of [14, 141, 151, 194, 195] for polycrystalline dysprosium.

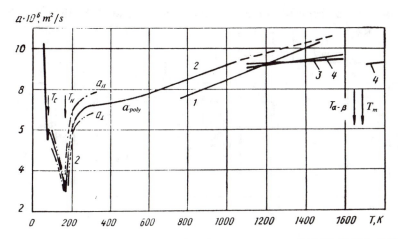

Figure 148 The diffusivity of dysprosium (a) vs. temperature: *1*) [140]; *2*) [39]; *3*) [141]; *4*) [150].

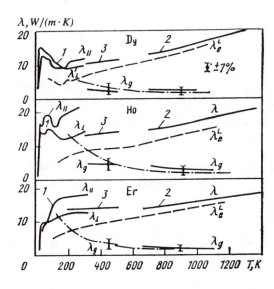

Figure 149 The thermal conductivity of dysprosium, holmium and erbium vs. temperature: *1*) [194], for single crystals in the direction of the c-axis ($\lambda_{\parallel}$), and perpendicular to it ($\lambda_{\perp}$); *2*) [140]; *3*) [48], for polycrystalline samples; $\lambda_e = \lambda_e^L = L_0 T/\rho$; λ_g is the estimate of [140].

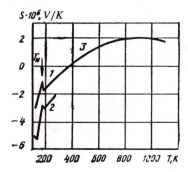

Figure 150 The absolute thermoelectric power of dysprosium (S) vs. temperature: *1*) [151]; *2*) Legvold, et al. [151].

Table 25 The kinetic properties of dysprosium [14, 48, 140, 151, 194, 195]

T, K	$\rho \cdot 10^{8}$, $\Omega \cdot$ m		$a \cdot 10^{6}$, m²/s	λ, W/(m·K)	T, K	$\rho \cdot 10^{8}$, $\Omega \cdot$ m		$a \cdot 10^{6}$, m²/s	λ, W/(m·K)
	poly	‖ c/⊥c	poly	poly		poly	‖ c/⊥c	poly	poly
100	—	52/50	—	—	800	152	—	8.5	14
200	88	70/94	6	10*	900	161	—	8.9	15
300	100	75/109	7.2	11	1000	169	—	9.2	16.5
400	110	—	7.3	11.5	1200	181	—	9.8*	18
500	120	—	7.5	12	1400	192	—	10.3	—
600	130	—	7.8	12.5*	1600	202	—	—	—
700	142	—	8.2	13					

*Interpolated values.

value decreases [151]. The kinetic properties of dysprosium are given in Table 25.

HOLMIUM

Below 1701 K, holmium crystalizes into an hcp structure with the lattice parameters a = 0.35773 nm and c = 0.56158 nm at 293 K [74]; and, above 1701 K, it has a bcc structure with a = 0.396 nm. At 131.6 K, holmium transforms from the paramagnetic into the antiferromagnetic state; and, at 17.5 K, it undergoes a ferromagnetic reordering.

The electronic structure of holmium is similar to that of gadolinium and the metals following it in the periodic table [41]. It is intermediate between the structures of gadolinium and lutetium.

The elastic constants of holmium ($c_{ij} \cdot 10^{-10}$, Pa) for the temperature range from 4.2 to 300 K are given below [193].

T, K	4.2	30	60	70	80	100	120
c_{11}	0.809	0.813	0.807	0.804	0.800	0.796	0.793
c_{33}	0.801	0.808	0.797	0.792	0.787	0.775	0.767
c_{44}	0.290	0.290	0.288	0.287	0.285	0.282	0.276
c_{12}	0.259	0.260	0.255	0.253	0.251	0.248	0.246
c_{13}	0.199	0.186	0.192	0.195	0.199	0.207	0.209

T, K	130	140	170	200	250	300
c_{11}	0.791	0.790	0.785	0.780	0.770	0.761
c_{33}	0.779	0.788	0.787	0.785	0.781	0.776
c_{44}	0.272	0.270	0.268	0.265	0.261	0.257
c_{42}	0.245	0.246	0.245	0.246	0.247	0.248
c_{13}	0.218	0.210	0.207	0.207	0.206	0.206

The difference of ±5% in the values of c_{ij}, given in the other studies [197, 198], leads to a 1.5 times difference in volume compressibility.

At $T_N = 132$ K, a helical structure with a spiral c-axis is formed. It is accompanied by the λ-type anomaly in the elastic constant c_{33}, and by a change in the slope of $c_{44}(T)$. At $T_c = 20$ K, holmium transforms to a ferromagnetic phase where the magnetic moment has one component in the direction of the hexagonal axis and another in the basal plane. This process is accompanied by anomalies in all of the elastic constants and by a sharp peak in the volume compressibility curve. The nature of the wide maximum in the compressibility curve and changes in the slopes of the curves for c_{11}, c_{12}, c_{13} and c_{66} near 80 K are probably due to changes in the magnetic order similar to those in gadolinium at 224 K and erbium at 53 K. The behavior of the moduli and volume compressibility of polycrystalline holmium has distinctive features at T_N, T_c at 80 K (Fig. 151).

The temperature dependence of the electrical resistivity of holmium is similar to that of dysprosium and terbium (Fig. 152). At high and moderate temperatures, the available data for polycrystalline holmium are consistent within 10%. The data extrapolated below 300 K agree within 3% with the mean data, obtained for single crystals. The data shown in Fig. 152 are applicable to metals of 99.2–99.9% purity with a relative residual resistance of $r = 15$ [140] and 20–30 [194].

The magnetic contribution, calculated from the average derivative of $\rho(T)$ with respect to temperature in the interval from 200 to 100 K, amounts to about 50% of the total resistivity $\rho_m = \rho - \rho_i - \rho_{e-ph}$, where $\rho_{e-ph} = AT$ [186]. At $T = 300$ K, the electrical resistivity from the scattering by disordered spins is $\rho_{e-s} = 24$ μΩ·cm. At high temperatures, a considerable decrease in $\partial\rho/\partial T$ could cause the decrease in the value of ρ_m ($\rho_{e-s} = 10$ μΩ·cm).

Figure 153 and Table 26 give the diffusivity of single crystal and polycrystalline holmium. At high temperatures, the difference between the data for polycrystalline holmium, obtained by various methods ([140] and [141, 150]), does not exceed 10%.

The results of the thermal conductivity investigations, given in Fig. 149, show that the electronic component of holmium represents the major part of its total thermal conductivity and determines its temperature dependence.

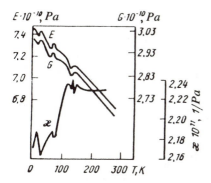

Figure 151 Young's (E), shear (G) moduli and volume compressibility (κ) of polycrystalline holmium vs. temperature [184].

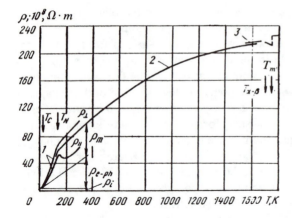

Figure 152 The electrical resistivity of holmium (ρ) vs. temperature: *1*) [188], for a single crystal in the direction of the *c*-axis ($\rho_{\parallel}$) and perpendicular to it ($\rho_{\perp}$); *2*) the average data from [140, 151, 150, 188, 195]; *3*) [195].

The temperature dependence of the thermoelectric power has a wide maximum at 800–1000 K [151] and anomalies near the points of magnetic disordering (Fig. 154).

The Hall coefficient of holmium has a temperature dependence [199] which is similar to that of the heavy rare earth metals preceeding it in the periodic table (Fig. 155). The nature of the dependence near 500 K, however, should be further investigated.

ERBIUM

At low and moderate temperatures, erbium, like the other heavy rare earth metals, has an hcp structure with the lattice parameters $a = 0.35588$ nm and $c = 0.55874$ nm at 293 K. It is assumed that, near the melting point, the hcp structure transforms into a bcc structure (the transformation temperature being 1643 K) [39]. Below 84 K, erbium transforms into an antiferromagnetic state, and, below 19.9 K, it has a conical ferromagnetic structure. The electronic structure of erbium is closer, in nature, to lutetium than to yttrium or gadolinium [41].

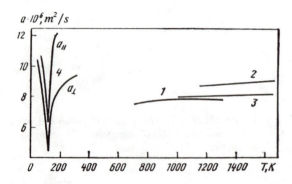

Figure 153 The diffusivity of holmium (*a*) vs. temperature: *1*) [140]; *2*) [148]; *3*) [150]; *4*) [39].

Table 26 The kinetic properties of holmium [39, 140, 195, 199]

T, K	$\rho \cdot 10^8$, $\Omega \cdot m$		$a \cdot 10^6$, m²/s	λ, W/(m·K)		T, K	$\rho \cdot 10^8$, $\Omega \cdot m$		$a \cdot 10^6$, m²/s	λ, W/(m·K)	
	poly	∥c/⊥c	poly	poly	∥c/⊥c		poly	∥c/⊥c	poly	poly	∥c/⊥c
100	—	40/48	—	—	16.0/14.0	800	162	—	7.9*	15.0	—
200	71	48/81	—	—	19.5/12.0	900	172	—	7.9*	15.5	—
300	90	62/100	—	13.0	—/13.5	1000	181	—	8.0	16.5*	—
400	106	—	—	13.5	—	1200	194	—	8.0	17.5*	—
500	123	—	—	14.0	—	1400	203	—	8.1	19.0*	—
600	136	—	—	14.0	—	1600	210	—	8.15	—	—
700	149	—	7.7*	14.5	—						

*The data need further investigation.

The elastic constants of hexagonal erbium ($c_{ij} \cdot 10^{-10}$, Pa) are given below [83].

T, K ...	0	20	40	55	60	80	85
c_{11}	8.700	8.910	8.496	8.873	8.859	8.787	8.771
c_{33}	8.120	8.387	8.335	8.418	8.529	8.502	8.491
c_{44}	2.760	2.826	2.912	2.916	2.916	2.884	2.871
c_{12}	—	—	—	2.985	2.925	2.916	2.913
c_{13}	2.312	2.340	2.390	2.212	2.214	2.928	2.213

T, K ...	100	140	180	220	260	300
c_{11}	8.735	8.668	8.596	2.518	8.444	8.367
c_{33}	8.598	8.580	8.562	8.525	8.483	8.445
c_{44}	2.858	2.846	2.825	2.801	2.777	2.753
c_{12}	2.907	2.911	2.915	2.920	2.925	2.929
c_{13}	2.220	2.216	2.214	2.215	2.217	2.222

Sharp changes in the elastic constants are observed near the temperatures of the magnetic phase transformations ($T_N = 85$ K, $T_r = 53$ K, $T_c = 20$ K). The restructuring of the antiferromagnetic order at $T_r = 53$ K is accompanied by struc-

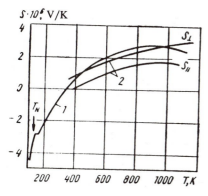

Figure 154 The absolute thermoelectric power of holmium [151] (S) vs. temperature: *1*) for a polycrystalline sample; *2*) for a single crystal in the direction of the *c*-axis ($S_∥$), and perpendicular to it ($S_⊥$).

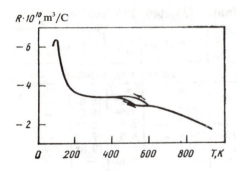

$R \cdot 10^{10}, m^3/C$

Figure 155 The Hall coefficient of holmium (R) vs. temperature [199].

tural distortions. This leads to the softening of the elastic moduli E and G in polycrystalline erbium. The dependences $G(T)$, $E(T)$ and $\varkappa(T)$ for high 99.9% purity erbium are shown in Fig. 156.

The temperature dependence of the electrical resistivity of polycrystalline erbium is nonlinear with $\partial^2 \rho / \partial T^2 < 0$ (Fig. 157). The data of [140, 151, 171] reproduce well the shape of the dependence of $\rho(T)$, and the values agree within 10%. The electron-phonon contribution was isolated from the average temperature coefficient of resistivity ($\rho_{e-ph} = AT$, where $A = \partial \rho / \partial T$). Its magnitude is close to that of the magnetic contribution ρ_m.

Figure 158 shows the diffusivity of polycrystalline and single crystal erbium. The curves are characterized by sharp minima near the magnetic transition points. Estimates in [39], based on thermal conductivity and specific heat data, agree within 10% with the direct measurements of polycrystalline erbium; the relative residual resistance $r = 15.5$.

At 200 K, the lattice and electronic components of the thermal conductivity of erbium are approximately equal. They were estimated in the same manner as for the other rare earth metals of the gadolinium subgroup. At higher temperatures, however, the value of λ_g decreases rapidly; and, at 800–1000 K the thermal conductivity is predominantly electronic in nature.

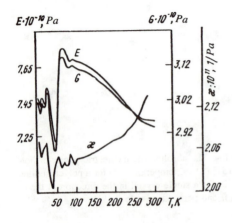

$E \cdot 10^{-10}, Pa$ $6 \cdot 10^{-10}, Pa$

Figure 156 Young's modulus (E), shear modulus (G) and volume compressibility $\varkappa(T)$ of polycrystalline erbium vs. temperature [184].

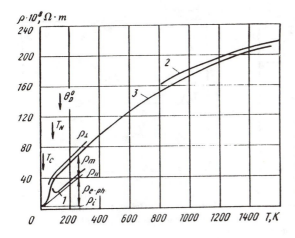

Figure 157 The electrical resistivity of erbium (ρ) vs. temperature: *1*) [14], for single crystal erbium in the direction of the *c*-axis ($\rho_\parallel$) and perpendicular to it ($\rho_\perp$); *2*) [171], for polycrystalline erbium; *3*) the average data of [14, 140, 150, 151, 171] for polycrystalline erbium.

The thermoelectric power of erbium increases with temperature, although towards 1000 K its temperature coefficient decreases substantially (Fig. 159). Like for other heavy rare earth metals, the Hall coefficient of erbium is negative, and its absolute value decreases. This is typical for metals with electronic conductivity (Fig. 160).

Table 27 summarizes the kinetic properties of erbium.

THULIUM

Up to temperatures near the melting point, thulium has an hcp structure with the lattice parameters $a = 0.35375$ nm and $c = 0.55546$ nm at 293 K [74]. It was suggested that the hcp structure can transform into a bcc structure [39] near the melting point. At low temperatures, thulium has a ferromagnetic structure which, above 22 K [39], transforms into a sinusoidal antiferromagnetic structure with the Neel point at $T_N = 55$ K.

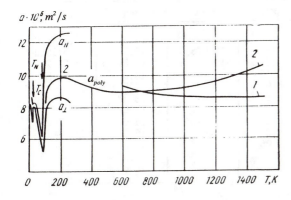

Figure 158 The diffusivity of erbium (*a*) vs. temperature: *1*) for polycrystalline erbium [140]; *2*) [39].

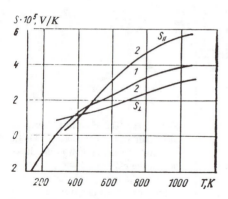

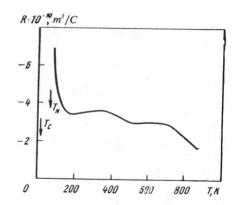

Figure 159 The absolute thermoelectric power of erbium (S) vs. temperature: 1) [151], for polycrystalline erbium; 2) [151], for single crystal erbium in the direction of the c-axis ($S_{\parallel}$) and perpendicular to it ($S_{\perp}$).

Figure 160 The Hall coefficient of polycrystalline erbium vs. temperature [199].

The electronic structure of thulium is similar to that of other heavy rare earth metals [41].

Figure 161 shows the temperature dependences of Young's and shear moduli for polycrystalline thulium [200]. The samples were of 99.9% purity and had a density of $d = (9.288 \pm 0.003)$ g/cm^3 at 298 K.

The absolute error in the determination of E and G is ~0.4%. The increase in moduli with the decrease in temperature is unusual. The slopes of the curves of $E(T)$ and $G(T)$ decrease sharply long before the temperature of magnetic ordering. At $T_N = 55$ K, minima that are typical for the magnetic transitions were observed, and, in the antiferromagnetic phase, the hardness of the lattice increases

Table 27 The kinetic properties of erbium [14, 150, 140, 151, 171]

T, K	$\rho \cdot 10^8$, $\Omega \cdot$m		$a \cdot 10^0$, m/s		λ, W/(m·K)	
	poly	$\parallel c / \perp c$	poly	$\parallel c / \perp c$	poly	$\parallel c / \perp c$
100	—	—	—	—	13.0	16.0/11.0
200	60	37/65	9.8	12.5/8.6	13.5	17.5/12.0
300	80	50/84	9.6	12.4/7.8	13.5	17.5/12.5
400	96	—	9.15	—	13.5	—
500	111	—	8.9	—	13.5	—
600	126	—	8.8	—	13.5*	—
700	140	—	8.8	—	14.2	—
800	152	—	8.9	—	14.9	—
900	162	—	9.0	—	15.5	—
1000	172	—	9.05	—	16.2	—
1200	190	—	9.4	—	17.5	—
1400	204	—	10.0	—	19.0	—

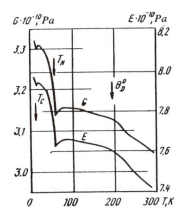

Figure 161 Young's (E) and shear (G) moduli of polycrystalline thulium vs. temperature [200].

substantially. The cusps in the curves of $E(T)$ and $G(T)$ near 38 K are associated with the transition into a new antiphase domain structure of a ferromagnetic type with the moments parallel to the c-axis. The phase transition, accompanied by a reorientation of ferrimagnetic planes which become parallel to the c-axis, is characterized by minima in both moduli at $T_c = 22$ K.

The temperature dependence of the electrical resistivity of thulium is shown in Fig. 162. Below 300 K, the data were obtained for single crystals of thulium with relative residual resistance of $r = 10 \div 20$ [201]; at higher temperatures, samples of polycrystalline thulium of about the same purity were used in [151] and [140]. The uncertainty in these data is ≈ 3–4%. The data of [140] for 300 K agree, within 5%, with the average data of [201], $\rho_{av} = 2/3\rho_\perp + 1/3\rho_\parallel$. The separation of contributions was made, as before, from the average slope of the electrical resistivity curve in the interval 150–800 K (Fig. 162). At moderate temperatures, the phonon and magnetic contributions are similar in magnitude. However, since the dependence of $\rho(T)$ is highly nonlinear above 1000 K, then, possibly, the value of ρ_m decreases, and ρ_{e-ph} deviates from the linear dependence.

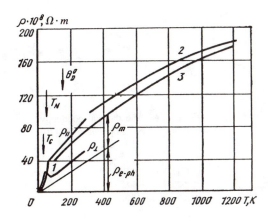

Figure 162 The electrical resistivity of thulium (ρ) vs. temperature: *1*) [201], for a single crystal; *2*) [140], for a polycrystalline sample; *3*) the average dependence of $\rho(T)$ for polycrystalline samples [140, 151, 201].

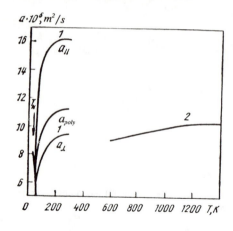

The diffusivity of thulium was calculated [39] from the thermal conductivity data (Fig. 163). The high temperature data [140] for polycrystalline samples with $r \approx 15$ have an uncertainty of 4%.

Figure 164 and Table 28 give the thermal conductivity and its components [140, 201]. The extrapolated low-temperature data of [201] for the direction perpendicular to the hexagonal axis agree, within 10%, with the data of [140]. At temperatures below 300 K, the electronic and lattice contributions to the thermal conductivity are similar in magnitude. However, above 300 K, the electronic component is dominant, and the difference between $\lambda_e^g = \lambda - \lambda_g$ and $\lambda_e^L = L_0 T/\rho$ becomes unnoticeable.

The temperature dependence of the thermoelectric power [151, 201] (Fig. 165) has anomalies near the low-temperature phase transition points. At moderate temperatures, it has a wide maximum similar to those in the other rare earth metals. The absolute value of the Hall coefficient of thulium [202] decreases with increases in temperature (Fig. 166).

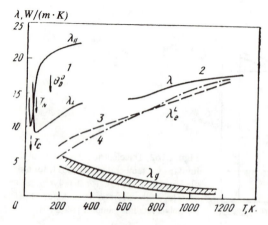

Figure 164 The thermal conductivity of thulium (λ) vs. temperature: *1*) [201], for a single crystal in the direction of the *c*-axis ($\lambda_\parallel$), and perpendicular to it ($\lambda_\perp$); *2*) [140], for a polycrystalline sample; *3*) $\lambda_e^L = L_0 T/\rho$; *4*) $\lambda_e^g = \lambda - \lambda_g$; λ_g is the estimate from [140].

Table 28 The kinetic properties of thulium [14, 140, 151, 201]

T, K	$\rho \cdot 10^8$, $\Omega \cdot m$		$a \cdot 10^6$, m²/s	λ, W/(m·K)		T, K	$\rho \cdot 10^8$, $\Omega \cdot m$		$a \cdot 10^6$, m²/s	λ, W/(m·K)	
	poly	$\|c/\perp c$	poly	$\|c/\perp c$	poly		poly	$\|c/\perp c$	poly	$\|c/\perp c$	poly
100	42	49/24	—	19.5//10	—	700	128	—	9.25	—	15
200	61	70/38	—	21.2//11.8	—	800	137	—	9.45	—	15.7
300	78	90/52	—	22//13.3	—	900	147	—	9.70	—	16.5
400	93	—	—	—	—	1000	157	—	9.90	—	17.1
500	104	—	—	—	—	1200	172	—	10.10	—	18.0
600	116	—	9.0	—	14.5	1400	—	—	10.15	—	—

YTTERBIUM

The information about the crystalline structure of ytterbium at room temperature is contradictory. Study [174] states that, at low temperatures, the γ-phase is stable, has an hcp structure and the parameters $a = 0.38799$ nm, $c = 0.63859$ nm at 296 K. In the vicinity of room temperature, this phase transforms into an fcc β-phase with $a = 0.54862$ nm at 293 K; the transformation is of a martensitic type. Finally, at 1965 K, the fcc phase transforms into a bcc γ-phase which $a = 0.444$ nm. Some sources, for example [74], cast doubt on the existence of the hcp modification. A possible reason for this disagreement is the influence of gaseous impurities.

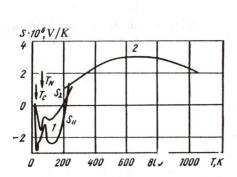

Figure 165 The absolute thermoelectric power of thulium vs. temperature: *1*) [201], for a single crystal in the direction of the c-axis ($S_{\|}$), and perpendicular to it ($S_{\perp}$); *2*) the data of [151] for polycrystalline sample.

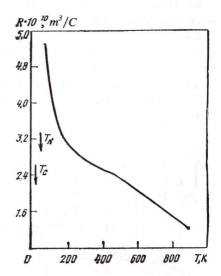

Figure 166 The Hall coefficient of polycrystalline thulium (R) vs. temperature [202].

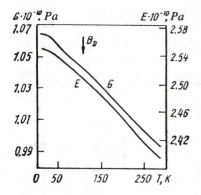

Figure 167 Young's (E) and shear (G) moduli of ytterbium vs. temperature [200].

The electronic structure of ytterbium is quite different from that of the other rare earth metals, and it is probably similar to the structure of europium [41]. The reason for this is that ytterbium is a divalent metal as its 4f-shell is full. Its coefficient of electronic heat capacity $\gamma = 2.92 \cdot 10^{-3}$ J/(mol·K^2) is substantially smaller than the value of $\gamma = 1 \cdot 10^{-2}$ J/(mol·K^2), which is the typical value for the majority of the other rare earth metals. This points to a smaller density of states at the Fermi level.

The Fermi surface of the hcp ytterbium should resemble that of cadmium and zinc [41].

Figure 167 shows the temperature dependence of Young's and shear moduli of polycrystalline ytterbium [200]. The purity of samples was 99.9%; the density, $(6.991 \pm 0.003) \cdot 10^3$ kg/m^3; the uncertainty in the values of G and E was ~0.4%.

The data on the temperature dependence of the electrical resistivity of ytterbium given in Fig. 168 are quite contradictory. The data of [140, 151] and some other results, discussed in these studies, point to the existence of a transition at 500–600 K, while the data of [151] point to a transition at 300–400 K. These transitions are accompanied by a hysteresis in $\rho(T)$. The data of [140, 151, 203] are given for ytterbium of $\approx$99.5% purity, with $r = 8$ [140]. In [179], a purer sample of ytterbium was used (see Fig. 168), and no transition was found below 1020 K. The transition detected at 1020 K was probably the fcc-bcc transformation.

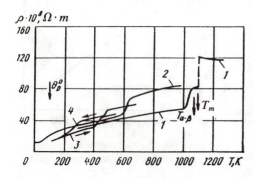

Figure 168 The electrical resistivity of ytterbium (ρ) vs. temperature: 1) [179]; 2) [140]; 3) [151]; 4) [14].

The information on the diffusivity of ytterbium below 500 K (Fig. 169) was obtained from the thermal conductivity data of [39]. At higher temperatures, the results of measurements are given for polycrystalline samples with $r = 8$ [140].

Figure 170 shows the thermal conductivity of ytterbium which is mainly of an electronic nature. Generally speaking, ytterbium has a much higher thermal conductivity than the other rare earth metals.

The values for the thermoelectric power of ytterbium are much higher than those of the other heavy rare earth metals and resemble those of europium (Fig. 171).

At room temperature, the Hall coefficient of ytterbium is about $10 \cdot 10^{-10}$ m^3/C [199], and decreases with increases in temperature. Its temperature dependence is complex and accompanied by a hysteresis, which requires a further investigation.

The kinetic properties of ytterbium are summarized below [14, 140, 179].

T, K	100	200	300	400	500	600	700	800	900	1000
$\rho \cdot 10^8$, $\Omega \cdot$m	20	30	33	36	40	44	47	50	52	54
λ, W/(m$\cdot$K)	—	39.0	36.0	34.5	33.5	—	—	—	—	—

LUTETIUM

Up to temperatures near the melting point, lutetium has an hcp structure with the lattice parameters $a = 0.35031$ nm and $c = 0.55509$ nm at 293 K. Near the melting point, the hcp structure probably transforms into a bcc structure [39, 74]. There is no information on the low-temperature magnetic phases for lutetium. The energy band structure and the Fermi surface of lutetium are similar to those of yttrium.

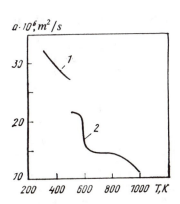

Figure 169 The diffusivity of ytterbium (*a*) vs. temperature: *1*) [140]; *2*) [39].

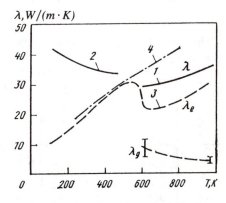

Figure 170 The thermal conductivity of ytterbium (λ) vs. temperature: *1*) [140]; *2*) [48]; *3*) $\lambda_e^L = L_0 T/\rho$, where ρ is from [179]; λ_g is the calculation of [140].

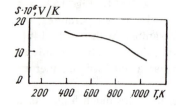

Figure 171 The thermoelectric power of ytterbium (S) vs. temperature [203].

The elastic constants of single crystal lutetium have been measured using two specimens. The first was purified by distillation, the second, by sublimation [204]. The values of c_{ij} differ by no more than 0.1%. For the temperature range from 4.2 to 300 K, no unusual feature in the behavior of the elastic moduli was observed. The elastic constants of the specimen, prepared by sublimation, have been calculated by taking into account its length. These data are presented below ($c_{ij} \cdot 10^{-10}$, Pa) [204].

T, K . . .	0	4.2	77.2	147.4	198.0	247.4	300.1
c_{11}	9.104	9.104	9.025	8.896	8.806	8.715	8.623
c_{33}	8.401	8.401	8.335	8.247	8.183	8.129	8.086
c_{44}	2.908	2.908	2.849	2.774	2.734	2.705	2.679
c_{66}	2.956	2.956	2.912	2.850	2.803	2.757	2.710
c_{12}	3.19	3.19	3.20	3.20	3.20	3.20	3.20
c_{13}	2.88	2.88	2.83	2.81	2.78	2.74	2.80

Table 29 summarizes the results of a systematic study of the elastic constants of all lanthanides, except the short-lived promethium. Samples of all metals were prepared through melting in an arc furnace and had purities of 99.80–99.85% by mass. The behavior of the elastic properties of lanthanides is consistently related to the number of electrons in the 4f-shell. Europium and ytterbium have the small-

Table 29 The elastic characteristics of polycrystalline rare earth metals at 300 K [205]

Metal	$K \cdot 10^{-10}$, Pa	$E \cdot 10^{-10}$, Pa	$g \cdot 10^{-10}$, Pa	μ	Metal	$K \cdot 10^{-10}$, Pa	$E \cdot 10^{-10}$, Pa	$g \cdot 10^{-10}$, Pa	μ
La	2.84	3.52	1.36	0.29	Tb	4.20	5.80	2.29	0.27
Ce	2.11	3.18	1.28	0.25	Dy	4.68	6.80	2.70	0.27
Pr	2.95	4.25	1.69	0.26	Ho	4.85	7.47	3.01	0.24
Nd	3.50	4.73	1.85	0.28	Er	5.78	7.05	2.67	0.30
Sm	3.63	4.98	1.99	0.26	Tm	4.51	6.12	2.38	0.28
Eu	1.02	1.56	0.62	0.25	Yb	1.41	1.92	0.75	0.27
Gd	4.14	5.87	2.33	0.26	Lu	6.71	6.94	2.55	0.33

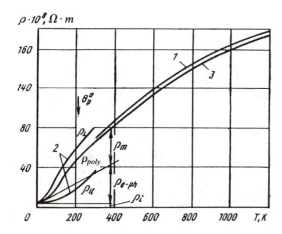

Figure 172 The electrical resistivity of lutetium (ρ) vs. temperature: *1*) [140]; 2) [194]; *3*) the average data of [140, 151, 194, 150].

est values of the moduli; they also manifest a specific valence of 2+, not the valence of 3+ found in the other rare earth metals [205].

The electrical resistivity of single crystals of lutetium with a relative residual resistence of $r \approx 10 \div 15$ is given in [194]. For polycrystalline lutetium of similar purity it is given in [140, 151]. The results of [140] and [151] agree among themselves, and with the average data of [194], within approximately ±5%.

The phonon component of the electrical resistivity of lutetium was isolated on the basis of that of yttrium (Fig. 172) with a correction for the difference in the Debye temperatures. The difference $\Delta\rho = \rho - \rho_i - \rho_{e-ph}$ is probably due to the same scattering mechanisms as in yttrium, and its temperature dependence is substantially nonlinear. Above 600–900 K, the values of both ρ and ρ_{e-ph} are nonlinear.

The diffusivity of lutetium, at high temperatures, was investigated in [140, 150] using a material of 99.5–99.8% purity. These data agree, within 10%, among themselves and with the results calculated for polycrystalline lutetium [39] (Fig. 173).

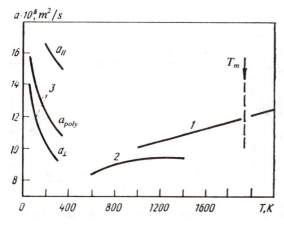

Figure 173 The diffusivity of lutetium (*a*) vs. temperature: *1*) [150]; 2) [140]; *3*) [39].

$\lambda, W/(m \cdot K)$

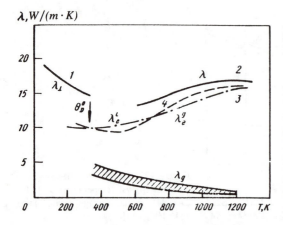

Figure 174 The thermal conductivity of lutetium (λ) vs. temperature: *1*) [194], for the *b*-axis (in the direction of the *c*-axis $\lambda_\parallel$ = 28, 26 and 24 W/(m·K) at 100, 200 and 300 K, respectively); *2*) [140]; *3*) $\lambda_e^g = \lambda - \lambda_g$; *4*) $\lambda_e^L = L_0 T/\rho$; λ_g is the calculation of [140].

Above 600 K, the thermal conductivity of lutetium is of an electronic nature, and the values of λ_e^g and λ_e^L differ by no more than 15%.

The thermoelectric power of lutetium is negative up to 1000 K, and $\partial^2 S/\partial T^2 < 0$, which is typical for the other heavy rare earth metals as well (Fig. 175). The temperature dependence of the Hall coefficient of lutetium (Fig. 176) is somewhat different from that of the other heavy rare earth metals. The kinetic properties of lutetium are listed in Table 30.

§4 TITANIUM, ZIRCONIUM, HAFNIUM

TITANIUM

Below 1155 K, titanium has an hcp structure with the lattice parameters a = 0.29511 nm and c = 0.46843 nm at 298 K. Above 1155 K, titanium has a bcc structure with a = 0.33065 nm at 1173 K [74].

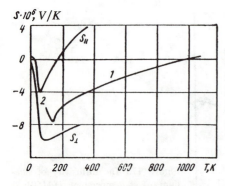

Figure 175 The absolute thermoelectric power of lutetium (*S*) vs. temperature: *1*) [194], for a polycrystal; *2*) [206].

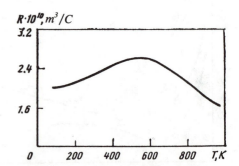

Figure 176 The Hall coefficient of lutetium (*R*) vs. temperature [202].

Table 30 The kinetic properties of lutetium [14, 39, 140, 151, 150]

T, K	$\rho \cdot 10^8$, $\Omega \cdot m$		$a \cdot 10^6$, m^2/s		λ, W/(m$\cdot$K)	
	poly	$\| c / \perp c$	poly	$\| c / \perp c$	poly	$\| c / \perp c$
100	20	8/30	14.4	—/11.7	—	28/18
200	46	18/57	12.2	16.5/10.0	—	26/16.5
300	65	36/80	11.!	15.4/9.1	16.5*	24/15
400	83	—	10.4*	—	14.5*	—
500	98	—	10.1*	—	13.5*	—
600	114	—	9.8*	—	13.0	—
700	127	—	9.7*	—	13.7	—
800	140	—	9.6	—	14.5	—
900	150	—	9.5	—	15.5	—
1000	160	—	9.6	—	16.0	—
1200	174	—	9.9	—	16.7	—
1400	—	—	10.2	—	17.1	—

*The data require further investigation.

The Fermi level of titanium [41] (Fig. 81) lies at the minimum of the curve of the density of states. At low temperatures, the electronic heat capacity of titanium is considerably smaller than that of its neighbours in the periodic table. Calculations of the Fermi surface of titanium produce quite contradictory results because small changes in the potential of the crystalline field lead to substantial changes in the details of the zone structure and Fermi surface.

Table 31 gives the elastic constants and volume compressibility of titanium for temperatures from 4 to 1100 K [207].

At room temperature, elastic moduli of titanium have the following values [135]: $E = 1.08 \cdot 10^{11}$ Pa; $G = 0.434 \cdot 10^{11}$ Pa; $K = 0.886 \cdot 10^{11}$ Pa.

Table 31 The elastic constants ($c_{ij} \cdot 10^{-11}$, Pa) and the volume compressibility ($\varkappa \cdot 10^{11}$, 1/Pa) of titanium [207]

T, K	c_{11}	c_{33}	c_{44}	c_{uu}	c_{13}	c_{12}	$\varkappa$
4	1.761	1.905	0.508	0.446	0.683	0.869	0.908
73	1.749	1.894	0.505	0.439	0.680	0.871	0.911
173	1.699	1.857	0.490	0.405	0.684	0.889	0.920
298	1.624	1.807	0.467	0.352	0.690	0.920	0.931
373	1.579	1.774	0.453	0.323	0.694	0.934	0.940
473	1.522	1.734	0.434	0.285	0.695	0.952	0.951
573	1.468	1.696	0.414	0.250	0.692	0.967	0.963
673	1.416	1.661	0.392	0.219	0.690	0.978	0.975
773	1.368	1.627	0.370	0.191	0.688	0.985	0.988
873	1.322	1.593	0.348	0.166	0.688	0.991	1.001
973	1.276	1.560	0.326	0.142	—	0.993	—
1073	1.231	1.529	0.307	0.118	—	0.996	—
1123	1.210	—	0.297	0.107	—	0.996	—
1153	1.197	—	0.291	0.102	—	0.996	—

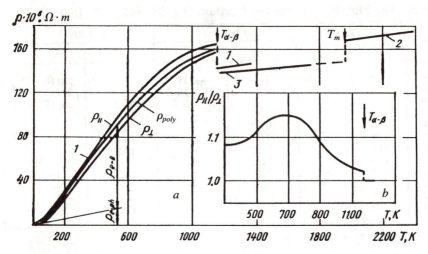

Figure 177 The temperature dependence of *a*) the electrical resistivity of titanium (ρ), *b*) the anisotropy in the electrical resistivity: *1*) [211, 140]; *2*) [210]; *3*) [208].

The coefficient of the shear anisotropy $A = c_{44}/c_{66}$ increases from 1.14, at 4 K, to 2.78, at 1123 K. The information on the electrical resistivity of polycrystalline and single crystal titanium is given in Fig. 177 and Table 32. It refers to a metal with the relative residual resistance of $r = 28$ [140]. Unlike the case of other hcp transition metals, for titanium $\rho_\parallel > \rho_\perp$. The anisotropy in resistivity

Table 32 The kinetic properties of titanium [39, 140, 208–212]

T, K	$\rho \cdot 10^8$, $\Omega \cdot m$			$a \cdot 10^6, m^2/s$, poly	λ, W/(m·K), poly
	poly	‖ *c*	⊥ *c*		
100	—	—	—	—	30
200	26.5	—	—	—	24
300	44.9	51.1	46.9	9.3	22
400	63.2	70.1	62.5	8.3	20
500	81.5	89.3	78.3	7.6	19
600	99.3	108.8	94.3	7.3	18.5
700	116	126.0	109.0	7.1	18.5
800	131	140.5	124.2	7.0	18.5
900	143	152.0	137.3	6.9	18.5
1000	152	157.5	147.5	6.85	19
1200	147	147.8	146.5	8.0	21.5
1400	151	—	—	8.3	24
1600	156	—	—	8.8	26
1800	160	—	—	9.0	27.5
1900	162	—	—	9.0*	—
2000	—	—	—	10.0*	—

*The data need verification.

reaches a maximum ($\approx$15%) at 600–700 K, but, at the α-β transformation point, it does not exceed 3%. During the α-β transformation the resistivity decreases in a 'jump', the slope and the shape of which is highly dependent on the sample purity [140, 208] (impurities decrease the slope of the 'jump' and displace the transformation point). At high temperatures, the dependence of $\rho(T)$ for polycrystalline and single crystal samples is substantially nonlinear and $\partial^2\rho/\partial T^2 < 0$. In the β-region the dependence of $\rho(T)$ has a small slope, similar to that of the liquid state. The uncertainty in the values given in Table 32 is $\approx$3%. Above 1600 K, the $\rho(T)$ of titanium has not been sufficiently studied.

For titanium, the mechanisms of the electron scattering have not been sufficiently studied. Study [140] indicated that, at high temperatures, the s-d model of Mott is not suitable for explaining the temperature dependences of the kinetic coefficients because $N(\varepsilon)$ lies at a minimum, and $\partial^2\rho/\partial T^2 < 0$. This is in contrast to the neighbouring scandium and vanadium, which lie near the maximum of $N(\varepsilon)$ while having the same sign of $\partial^2\rho/\partial T^2$. Figure 177 also shows the electron-phonon contribution to the electrical resistivity which was obtained from the electrical resistivity of calcium, a close neighbour of titanium, by taking into account the difference in the Debye temperatures. The slope of the electron-phonon contribution is close to that of the $\rho(T)$ curve for β-Ti. The remaining part of the resistivity, which was labelled as ρ_{e-e}, is probably due to the s-d scattering. Study [140] indicated that the nonlinearity of $\rho(T)$ and the disappearance of the anisotropy in ρ may be explained by a decrease in the mean free path of electrons, approaching the interatomic distance. This possibility was suggested in [209], which discussed the specific features of the phonon spectrum and its influence on specific heat and electrical resistivity.

The diffusivity of polycrystalline titanium [140] with $r = 90$ jumps during the α-β transformation, the slope of the jump depends on the purity of the samples. The uncertainty in the data, given in Table 32 and Fig. 178, is $\approx$10%.

The thermal conductivity of titanium has a wide minimum in the region of 600–800 K. At higher temperatures, it has a positive temperature coefficient (Fig. 179). During the α-β transformation there is a small change in thermal conductivity. Above 600 K, the contribution of the lattice component to thermal con-

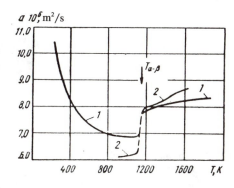

Figure 178 The diffusivity of titanium (a) vs. temperature: *1*) [39]; 2) [140].

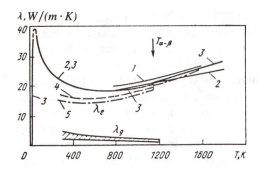

Figure 179 The thermal conductivity of titanium (λ) vs. temperature: *1*) [39]; *2*) [140]; *3*) the average of [39, 140, 211, 212]; *4*) $\lambda_e^g = \lambda - \lambda_g$; *5*) $\lambda_e^L = L_0 T/\rho$; λ_g is the calculation from [140].

ductivity is negligible considering a possible uncertainty of 10%. However, below about 400 K, $\lambda_e^g = \lambda - \lambda_g$ is somewhat greater than $\lambda_e^L = L_0 T/\rho$. This is probably due to the influence of the inelastic contributions to the *s-d* scattering.

The temperature dependence of the thermoelectric power of titanium also has a minimum near 700–900 K, and, in the region of the β-phase, its value is close to zero [55] (Fig. 180).

At room temperature, the Hall coefficient of titanium is equal to $-2.4 \cdot 10^{-10}$ m^3/C; its absolute value decreases with decreases in temperature to $-0.3 \cdot 10^{-10}$ m^3/C at 100 K [82].

ZIRCONIUM

At temperatures below 1135 K, zirconium has an hcp structure with the lattice parameters $a = 0.32312$ nm and $c = 0.51477$ nm at 298 K [74]. Above 1135 K it has a bcc structure with $a = 0.36090$ nm near the α-β transformation [82].

The Fermi level of zirconium, as in titanium, lies at the minimum of the density of electronic states (Fig. 87).

Study [41] described two, not entirely compatible, models of the Fermi surface of zirconium. It is known, however, that its first and second Brillouin zones are full, and that the third and fourth zones have a number of closed surfaces of holes. The Fermi surface of zirconium in the fifth and sixth zones, according to the data of Altmann and Bradley, is shown in Fig. 181 [41].

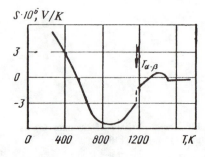

Figure 180 The absolute thermoelectric power of titanium (S) vs. temperature [55].

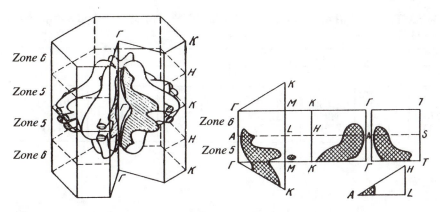

Figure 181 The Fermi surface of zirconium in the fifth and sixth zones (the electronic region is shaded) [41].

The elastic moduli of α-Zr were measured for a broad temperature range using the textured rod samples [213]. Changes in the values of Young's $E(T)$ and shear $G(T)$ moduli vs. orientation are shown in Fig. 182 from the direction parallel to the c-axis to the direction perpendicular to it. Table 33 gives the values of the moduli for several temperatures for the directions parallel and perpendicular to the c-axis.

Young's modulus of polycrystalline β-Zr containing about $210 \div 125$ ppm of oxygen was measured up to 1473 K (Fig. 183). The value of E has a linear temperature dependence, $\partial E / \partial T = 4.66 \cdot 10^7$ Pa/K.

Table 34 gives the elastic constants for single crystal zirconium for the range from 4 to 1100 K. The temperature dependences of c_{ij} have the following unusual features: 1) above 200 K, the temperature dependence of c_{66} and c_{44} has a positive curvature; 2) in the range from 4 to 1100 K the shear modulus c_{66} decreases more

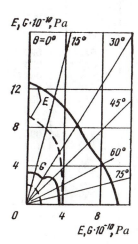

Figure 182 The elastic anisotropy in single crystals of α-Zr at 272 K (continuous line) and 1133 K (broken line); θ is the angle relative to the c-axis [213].

Table 33 The moduli of elasticity (E, G, $K \cdot 10^{-10}$, Pa) of textured samples of α-Zr [213]

T, K	‖ c			⊥ c		
	E	G	K	E	G	K
4.2	13.501	3.630	32.176	11.711	3.982	27.852
273	12.609	3.240	31.802	10.081	3.424	27.232
1133	9.215	2.010	29.706	3.840	1.447	24.411

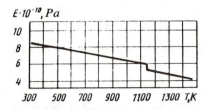

$E \cdot 10^{-10}$, Pa

Figure 183 Young's modulus of zirconium (E) vs. temperature [214].

(75%) than c_{44} (42%), the latter is usually more sensitive to temperature changes; 3) the anisotropy in the elastic compression constant c_{11} decreases by 32% and in c_{33} by 21%. Similar features are observed for titanium and hafnium.

The temperature dependence of the electrical resistivity of zirconium is given in Fig. 184 for a sample of about 99.9% purity according to the data compiled in [82] and [215]. In the β-phase region, the data require further investigation. The temperature dependence of the electrical resistivity of zirconium resembles

Table 34 The elastic constants ($c_{ij} \cdot 10^{-11}$, Pa) and the volume compressibility ($\varkappa \cdot 10^{11}$, 1/Pa) of Zirconium [207]

T, K	c_{11}	c_{33}	c_{44}	$c_{\omega i}$	c_{13}	c_{12}	$\varkappa$
4	1.554	1.725	0.363	0.441	0.646	0.672	1.029
73	1.542	1.716	0.358	0.432	0.648	0.678	1.030
173	1.495	1.687	0.342	0.401	0.648	0.692	1.042
298	1.434	1.648	0.320	0.953	0.653	0.728	1.050
373	1.396	1.623	0.308	0.325	0.654	0.746	1.058
473	1.347	1.591	0.295	0.290	0.654	0.767	1.069
573	1.301	1.559	0.282	0.257	0.657	0.786	1.078
673	1.257	1.526	0.270	0.229	0.659	0.799	1.090
773	1.214	1.493	0.257	0.201	0.659	0.812	1.102
873	1.168	1.460	0.245	0.175	0.659	0.819	1.118
973	1.127	1.425	0.232	0.149	0.657	0.828	1.132
1073	1.087	1.393	0.220	0.125	0.656	0.838	1.146
1105	—	1.382	—	0.117	—	0.840	—
1133	1.064	—	—	—	—	—	—

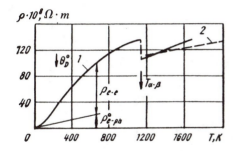

$\rho \cdot 10^{8}, \Omega \cdot m$

Figure 184 The electrical resistivity of zirconium (ρ) vs. temperature: *1)* [82]; *2)* [215].

that of titanium. The electron-phonon contribution from the scattering of *s*-electrons may be isolated using the resistivity of strontium, the neighbour of zirconium in the periodic table (see Fig. 184). As can be seen, the slope of this straight line does not differ by much from the slope of the curve of $\rho(T)$ for β-Zr. As in titanium, the difference $\Delta\rho = \rho - \rho_{e-ph}$ is probably due to the nature of the electron-electron *s-d* scattering; zirconium differs from strontium by two *d*-electrons.

Above 600 K, zirconium may also be affected by the fact that the mean free path of electrons approaches the interatomic distance [140].

Study [140] contains the diffusivity data for zirconium samples where $r = 40$. Near the α-β transformation the shape of the 'jump' depends strongly on the sample purity. The presence of impurities makes the 'jump' less steep and displaces it along the temperature scale. The uncertainty is $\approx 5\%$ for the α-region and $\approx 10\%$ for the β-region (Fig. 185). The kinetic properties of zirconium are summarized below [39, 48, 82, 140, 215, 216, 226].

T, K	200	300	400	500	600	700	800
$\rho \cdot 10^{8}$, $\Omega \cdot m$	26.8	43.5	60.2	76.3	91.2	103	113
$a \cdot 10^{6}$, m^2/s	14.1	12.7	11.5	10.75	10.2	10.0	10.0
λ, W/(m·K)	32	27	24.5	23	22	22	22

T, K	900	1000	1200	1400	1600	1800	2000
$\rho \cdot 10^{8}$, $\Omega \cdot m$	121	126	110	114	118	123	127
$a \cdot 10^{6}$, m^2/s	10.0	9.8	13.2	14.2	14.7	14.9	14.5
λ, W/(m·K)	22.5	23	26.5	29.5	33	36	—

The thermal conductivity of zirconium is given in Fig. 186. The curve of $\lambda(T)$ has a minimum near 500–800 K, following which it increases with temperature. The increase is caused by the electronic component of $\lambda(T)$. Below 800 K, the value of λ_e^g is greater than that of λ_e^L, therefore $L > L_0$, which may be the result of the *s-d* scattering. However, at higher temperatures, the influence of the band effects is insignificant, and the electron scattering becomes elastic. In the region of the β-phase, within the uncertainty limits of $\approx 10\%$, λ can be estimated to equal λ_e.

The temperature dependence of the thermoelectric power of zirconium, as that

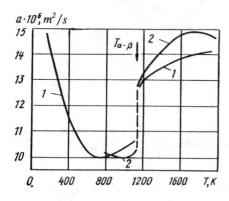

Figure 185 The diffusivity of zirconium (*a*) vs. temperature: *1*) [39]; *2*) [140].

of titanium, has a wide minimum at moderate temperatures, and in the region of the β-phase its values approach zero (Fig. 187) [55].

The Hall coefficient of zirconium (*r* = 38) changes from $2.15 \cdot 10^{-10}$ m³/C at 300 K to $3.15 \cdot 10^{-10}$ m³/C at 100 K.

HAFNIUM

At standard pressure, hafnium has an hcp structure, like titanium and zirconium, with the lattice parameters *a* = 0.31946 nm, *c* = 0.50511 nm at 293 K [74]. At 2030 ± 20 K the α-phase transforms into β-phase, the latter has a bcc structure with *a* = 0.3615 nm at 2053 K.

Unlike its neighbours in the periodic table, hafnium has a low density of the electronic states; in Fig. 87 the value of $N(\varepsilon)$ lies near the minimum. The Fermi surface of hafnium has not been calculated, it can be assumed to be similar to those of zirconium and titanium.

Table 35 lists the elastic constants of a hafnium sample, containing 4.1% (by

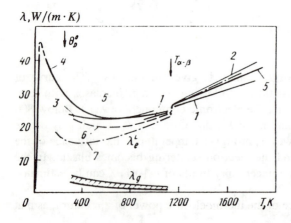

Figure 186 The temperature dependence of the thermal conductivity of zirconium (λ) vs. temperature: *1*) [140]; *2*) [226]; *3*) [48]; *4*) [216]; *5*) the average values; *6*) $\lambda_e^g = \lambda - \lambda_g$; *7*) $\lambda_e^L = L_0 T/\rho$; λ_g is calculated in [140].

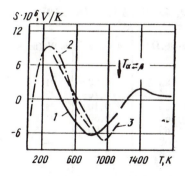

Figure 187 The absolute thermoelectric power of zirconium (*S*) vs. temperature: *1*) [55]; *2*) the data of Potter [55]; *3*) the data of Odensted [55].

mass) of Zr, 0.01% (by mass) of O_2 and $\approx 0.01\%$ (by mass) of other impurities [207]. As in titanium and zirconium, the hafnium constant c_{66} has a stronger temperature dependence than c_{44}. The coefficient of the shear anisotropy $A_{sh} = c_{44}/c_{66}$ is greater than one and increases with temperature.

The elastic moduli of polycrystalline hafnium have the following values at 293 K [135]: $E = 1.41 \cdot 10^{11}$ Pa; $G = 0.558 \cdot 10^{11}$ Pa; $K = 1.09 \cdot 10^{11}$ Pa.

Table 36 and Fig. 188 [215] show the temperature dependence of the electrical resistivity of polycrystalline hafnium prepared by the iodide method and containing up to 1.5% Zr. The information on the anisotropy of the electrical resistivity of hafnium is not available. By analogy with titanium, the anisotropy can be assumed to be 10% at moderate temperatures. The uncertainty in the data given in Fig. 188 and Table 36 is estimated as 10%. For the β-phase the electrical resistivity of hafnium requires further investigation.

The discussed dependence is essentially non-linear, and $\partial\rho/\partial T$ is close to zero at temperatures above 1600 K. The electron-phonon resistivity from the phonon scattering of the *s*-electrons can be estimated from the resistivity of barium, a close nontransition neighbour of hafnium (see Fig. 188). The remaining part of

Table 35 The elastic constants ($c_{ij} \cdot 10^{-11}$, Pa) and the volume compressibility ($\varkappa \cdot 10^{11}$, 1/Pa) of Hafnium [207]

T, K	c_{11}	c_{33}	c_{44}	c_{66}	c_{13}	c_{12}	$\varkappa$
4	1.901	2.044	0.600	0.578	0.655	0.745	0.903
73	1.891	2.035	0.595	0.572	0.658	0.747	0.904
123	1.875	2.022	0.588	0.562	0.659	0.751	—
173	1.859	2.008	0.579	0.551	0.660	0.757	0.911
223	1.842	1.993	0.571	0.539	0.661	0.765	—
273	1.822	1.977	0.562	0.526	0.661	0.770	0.918
298	1.811	1.969	0.557	0.520	0.661	0.772	—
373	—	—	—	0.500	—	—	—
473	—	—	—	0.472	—	—	—
573	—	—	—	0.443	—	—	—

Table 36 The kinetic properties of hafnium [39, 48, 82, 215, 218, 219]

T, K	$\rho \cdot 10^8$, $\Omega \cdot m$	$a \cdot 10^6$, m^2/s	λ, W/(m·K)	T, K	$\rho \cdot 10^8$, $\Omega \cdot m$	$a \cdot 10^6$, m^2/s	λ, W/(m·K)
200	26.7	—	25	1000	125.3	8.50	21
300	39.3	11.55*	23.5	1200	144.8	8.70	22
400	51.9	10.7*	22.5	1400	158	9.10	24
500	64.5	10.0*	22	1600	166	9.60	25.5
600	77.08	9.5*	21	1800	169	9.90	27.5
700	89.61	9.1*	21	2000	162	—	—
800	101.9	8.8*	20.5	2200	158	12.8	—
900	113.9	8.6*	20.5	2400	—	12.5	—

*The data require further investigation.

the resistivity $-\Delta\rho = \rho - \rho_{e-ph}$ has the same magnitude as ρ_{e-ph} and can be explained by the electron scattering typical for the transition metals. This may be electron-electron scattering, such as in titanium and zirconium, however the possibility of Mott's s-d band scattering cannot be excluded. The nonlinearity in $\rho(T)$ above 1000 K may be caused by the decrease in the mean free path of s-electrons, which approaches the interatomic distance [140].

The information on the diffusivity of hafnium is given in Table 36 and Fig. 189, it applies to polycrystalline hafnium, prepared by the iodide method and containing 0.3–1% Zr [140]. The uncertainty in the values is 10%. The shape of the diffusivity 'jump' near the α-β transformation strongly depends on the purity of the samples.

The temperature dependence of the thermal conductivity of polycrystalline hafnium has a wide minimum at moderate temperatures (Fig. 190). Above 800 K, the Lorenz number is close to the standard value of L_0, supporting the fact that the elastic processes of electron scattering are predominant. However, at low and moderate temperatures, $\lambda - \lambda_g > L_0T/\rho$, which may point to the presence of inelastic and band effects.

The temperature dependence of the thermoelectric power of hafnium [55] has

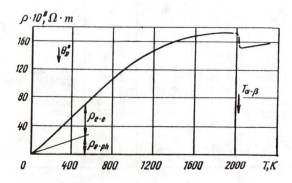

Figure 188 The electrical resistivity of hafnium (ρ) vs. temperature (the average of [82, 215, 218]).

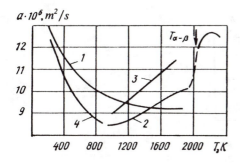

Figure 189 The diffusivity of hafnium (a) vs. temperature: 1) [39]; 2) [140]; 3) [218]; 4) [219].

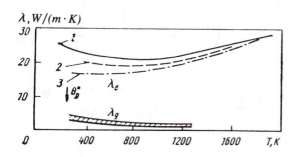

Figure 190 The thermal conductivity of hafnium (λ) vs. temperature: 1) the average of [48, 140, 215, 218, 219]; 2) $\lambda_e^g = \lambda - \lambda_g$; 3) $\lambda_e^L = L_0 T/\rho$; λ_g is according to the calculation of [140].

a wide minimum at moderate temperatures and is similar to those observed in zirconium and titanium (Fig. 191).

The Hall coefficient of hafnium changes monotonically from $+0.43 \cdot 10^{-10}$ m^3/C at 300 K to $0.19 \cdot 10^{-10}$ m^3/C [82, 100].

§5 VANADIUM, NIOBIUM, TANTALUM

The metals of the fifth subgroup, vanadium, niobium, and tantalum, are the classical refractory metals. Their crystalline and electronic structures, and the temperature dependences of their kinetic properties have many common features (see Fig. 192 and 193, where the elastic constants are provided).

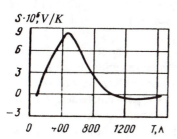

Figure 191 The absolute thermoelectric power of hafnium (S) vs. temperature [55].

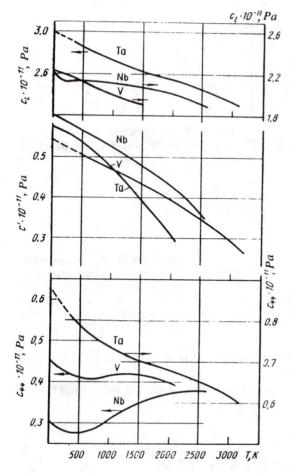

Figure 192 The elastic constants of vanadium, niobium, and tantalum vs. temperature [220–222].

VANADIUM

Vanadium has a bcc crystalline structure with $a = 0.30282$ nm at 303 K [39, 74].

The Fermi level of vanadium lies near the maximum of the density of the electronic states (see Fig. 81). The main features of this surface are: a region of holes in the second zone at the Γ point, a multiply-connected net of tubes of holes in the third zone in the ΓH direction (nicknamed 'jungle-gym'), and ellipsoidal pockets of holes in the third zone at the N point.

The data on the electrical resistivity of vanadium are compiled in [215] (Fig. 194). At 300 K, the average electrical resistivity of pure vanadium is 20.1 ± 0.5 $\mu\Omega \cdot$ cm (95% confidence interval). The electrical resistivity can be expressed as a function of the sample purity: $\rho(c) = 20.09 + 6.11\,c$ $\mu\Omega \cdot$ cm, where c is the total concentration of impurities in %(at.); ρ is the electrical resistivity, $\mu\Omega \cdot$ cm.

Study [215] mentioned investigations which discuss a possibility of a phase

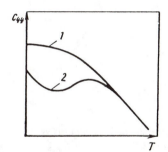

Figure 193 The shear elastic constant c_{44} as a function of temperature: *1*) the typical behavior; *2*) the anomalous behavior, specific for the 3*d*-, 4*d*-, and 5*d*-transition metals with a bcc structure.

transformation in vanadium at 190–230 K. The transformation is debatable because the data have not been reproduced. The recommended resistivity values of pure vanadium are summarized below (taking into account thermal expansion; the uncertainty is ~5%) [39, 140, 215, 226, 227].

T, K	200	300	400	500	600	700	800	900
$\rho \cdot 10^8$, $\Omega \cdot$m	12.7	20.1	27.1	33.8	40.1	46.1	51.9	57.4
$a \cdot 10^6$, m^2/s	12.1*	10.4	10.1	10.1	10.2	10.4	10.6	10.8

T, K	1000	1200	1400	1600	1800	2000	2100
$\rho \cdot 10^8$, $\Omega \cdot$m	62.7	72.7	82.1	91.1	99.9	108.8*	113.3*
$a \cdot 10^6$, m^2/s	11.2	11.6	11.5	11.5	11.4	11.2*	—

*The data need further investigation.

Study [140] suggested that unlike niobium and tantalum, vanadium has an additional scattering mechanism. Its contribution can be isolated, using the temperature dependence of the electrical resistivity of niobium (see Fig. 194).

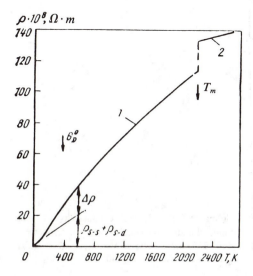

Figure 194 The electrical resistivity of vanadium vs. temperature: *1*) the data of [215]; *2*) the data of [201].

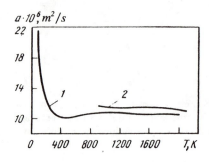

Figure 195 The diffusivity of vanadium (a) vs. temperature: 1) [39]; 2) [140].

This contribution represents a sum of the s-s and s-d contributions to the resistivity, where the s-d component is due to the s-d band scattering. The additional contribution could result from scattering of s-electrons by localized spin fluctuations [140].

The temperature dependence of diffusivity has a small wide minimum near 400–500 K, whereas, above 800 K, the diffusivity has a weak temperature dependence. Uncertainty in the data given in Fig. 195 is ≈5%.

Above 300 K, the temperature dependence of the thermal conductivity has a positive slope, and, above 1500 K, the slope is nearly zero and even negative (Fig. 196). Above 1200 K, the Lorenz number is close to L_0. However, in vanadi-

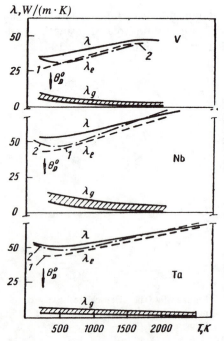

Figure 196 The thermal conductivity of vanadium, niobium and tantalum (λ) vs. temperature [140, 225, 226]: 1) $\lambda_e^g = \lambda - \lambda_g$; 2) $\lambda_e^L = L_0 T/\rho$; λ_g is the calculations of [140].

um and some of its alloys, at low and moderate temperatures, the Lorenz number may decrease, in comparison with L_0, as a result of inelastic electron scattering.

The temperature dependence of the thermoelectric power of vanadium has a positive slope which decreases at high temperatures (Fig. 197), so that, above 1000 K, a maximum of $S(T)$ is observed [55].

The Hall coefficient of vanadium is equal to $(0.76 \div 0.82) \cdot 10^{-10}$ m³/C at 300 K. When the temperature decreases, the temperature dependence of the Hall coefficient diminishes $R = 0.62 \cdot 10^{-10}$ m³/C at 20 K [82, 100].

NIOBIUM

Niobium has a bcc crystalline structure with the lattice parameter $a = 0.33005$ nm at 298 K [39, 74]. The Fermi level of niobium lies near the maximum of the density of electronic states (see Fig. 87).

The temperature dependence of the electrical resistivity of niobium is shown in Fig. 198. At 300 K, the influence of impurities on the electrical resistivity of niobium can be described by the following equation: $\rho_{300K} = 14.77 + 0.86\ c$, where c is the concentration of 'heavy' impurities in % (by mass). The confidence interval for the value of ρ_{300K} is about 2% at the 95% confidence level. The uncertainty does not change up to 2000 K, and, at 2500 K, the confidence interval becomes about 3%.

The resistivity values of niobium, corrected for the thermal expansions, are given below. The electrical resistivity of niobium, containing 0.2% impurities, was measured near the melting point using the exploded wire method [227]. The obtained values were $\rho_{sol} = 95.2$ μΩ·cm, $\rho_{liq} = 108.5$ μΩ·cm [227]. The non-linearity of the temperature dependence of the electrical resistivity is typical for the transition metals in which the Fermi levels lie near the maxima of the density

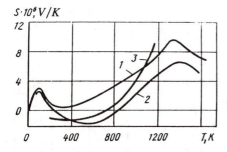

Figure 197 The absolute thermoelectric power of vanadium (S) vs. temperature [55]: *1*) the combined temperature dependence of the absolute thermoelectric power of vanadium, niobium and tantalum [55]; *2*) the thermoelectric power of single crystals of tantalum; *3*) [225].

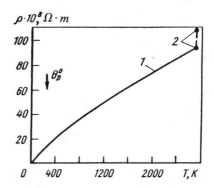

Figure 198 The electrical resistivity of niobium (ρ) vs. temperature: *1*) [226]; *2*) [227].

of the electronic states. Studies [140, 215] showed that, within the Mott *s-d* model, the band scattering can be related to the parameters of the curve of the density of states. Thus, there is a correlation between the temperature dependence of the electrical resistivity and magnetic susceptibility (see expressions (28) and (29)).

The diffusivity coefficient of niobium, like that of vanadium, has a wide minimum near room temperature, while for the 1000–2000 K interval its temperature dependence is weak. The uncertainty in the data, shown in Fig. 199 and given below, is ≈10%.

The temperature dependence of the thermal conductivity of niobium also has a wide minimum near room temperature and a positive slope above 500 K (Fig. 196). Above 500 K, the values of λ_e^g and λ_e^L coincide within the ≈10% uncertainty in the determination of λ. However, below 300 K, they are different ($\lambda_e^L > \lambda_e^g$) because of the band effects.

Above 800 K, the temperature coefficient of the thermoelectric power of niobium decreases, as does that of vanadium (Fig. 200).

The Hall coefficient of niobium is equal to $8 \cdot 10^{-10}$ m³/C at 273 K, and has a weak temperature dependence ($R = 0.92 \cdot 10^{-10}$ m³/C at 83 K and $R = 0.91 \cdot 10^{-10}$ m³/C at 873 K) [100].

The kinetic properties of niobium are summarized below [39, 48, 140, 215, 226].

T, K	100	200	300	400	500	600	700	800	900	1000
$\rho \cdot 10^8$, Ω·m	—	9.71	14.7	19.5	23.8	27.7	31.4	34.9	38.3	41.6
$a \cdot 10^6$, m²/s	34.5	24.5	23.7	23.5	23.5*	23.9*	23.9*	24.0*	24.2*	24.5*
λ, W/(m·K)	55.2	52.6	53.7	55.2	56.7	58.2	59.8	61.3	62.9	64.4

T, K	1200	1400	1600	1800	2000	2200	2400	2600	2800
$\rho \cdot 10^8$, Ω·m	47.9	54.0	60.0	65.9	71.8	77.6	83.3	89.0	—
$a \cdot 10^6$, m²/s	24.9*	25.0*	25.0*	25.0*	24.6*	24.0*	23.0*	21.7*	20.3*
λ, W/(m·K)	67.5	70.5	73.5	76.4	—	—	—	—	—

*The data needs revision.

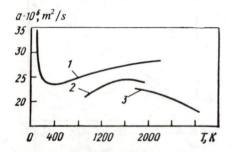

Figure 199 The diffusivity of niobium (*a*) vs. temperature: *1*) [39]; 2) the result of Weller [140]; *3*) [217].

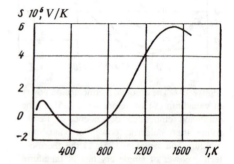

Figure 200 The absolute thermoelectric power of niobium (*S*) vs. temperature [55].

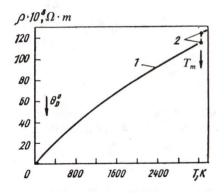

Figure 201 The electrical resistivity of tantalum (ρ) vs. temperature: *1*) [226]; *2*) [227].

TANTALUM

Similar to vanadium and niobium, tantalum has a bcc structure with the lattice parameter $a = 0.3297$ nm at 298 K.

The Fermi level of tantalum also lies near the maximum on the curve of the density of the electronic states (see Fig. 93) [41].

The data on the electrical resistivity of tantalum over a broad temperature interval are compiled in [215, 226]. At 300 K, the electrical resistivity of pure tantalum is $\rho = 13.4 \pm 0.3$ $\mu\Omega \cdot$cm, and the values of ρ can be expressed as: $\rho = 13.45 + 0.69$ c, where c is the concentration of the 'heavy' impurities, including carbon in % (by mass).

Figure 201 shows the temperature dependence of the electrical resistivity of tantalum. As in vanadium and niobium, it is nonlinear, with $\partial^2\rho/\partial T^2 < 0$, and the values of the electrical resistivity differ little from those of niobium. The combination of the *s-s* and *s-d* mechanisms of scattering probably takes place in tantalum as well [140]. The parameters of the electronic structure were calculated in [140] from the temperature dependences of the electrical resistivity ρ and the magnetic susceptibility $\varkappa$. The agreement in the sign and values of the derivatives of ρ and $\varkappa$, with respect to energy, was satisfactory. This indicates that, qualitatively, the temperature dependences of these two properties at high temperatures are of the same nature and are caused by the *s-d* scattering mechanism of Mott.

The diffusivity of tantalum for the temperature range 400–2000 K is practically constant (Fig. 202), but at higher temperatures it decreases [140, 228]. The uncertainty of these values is $\approx 5\%$.

The temperature dependence of the thermoelectric power of tantalum is similar to that of vanadium and niobium (see Fig. 197, 200): at moderate and high temperatures the value of $\partial S/\partial T$ decreases with increasing temperature.

The Hall coefficient of tantalum is equal to $0.971 \cdot 10^{-10}$ m^3/C at 273 K and it has a weak temperature dependence ($R = 0.975 \cdot 10^{-10}$ m^3/C at 83 K and $R = 0.967 \cdot 10^{-10}$ m^3/C at 773 K) [82, 100].

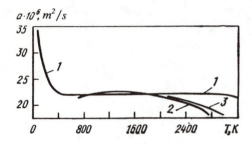

Figure 202 The diffusivity of tantalum (a) vs. temperature: _1_) [39]; _2_) [140]; _3_) [228].

The temperature dependence of the thermal conductivity of tantalum has a wide minimum at low temperatures and a positive temperature coefficient. Above 1000 K, the Lorenz number $L \to L_0$, and below 200 K $\lambda_e^L > \lambda_e^g$, which can be caused by the inelastic effects. In the interval 200–1000 K, the difference between λ_e^g and λ_e^L may be caused by the band effects [215, 225]. The uncertainty in the values of λ given below is $\approx 10\%$ up to 2500 K, and, at higher temperatures, the data require further investigation.

The kinetic properties of tantalum are given below [140, 215, 226].

T, K	200	300	400	500	600	700	800	900	1000	1200
$\rho \cdot 10^8$, $\Omega \cdot m$	8.76	13.4	18.0	22.5	26.9	31.2	35.3	39.5	43.5	51.3
$a \cdot 10^6$, m^2/s	26.9	23.7	22.1	22.0	22.0	22.0	22.0	22.0	22.0	22.0
λ, W/(m·K)	53.0	52.1	51.8	51.8	51.8	52.0	52.4	52.8	53.2	54.4

T, K	1400	1600	1800	2000	2200	2400	2600	2800	3000	3200
$\rho \cdot 10^8$, $\Omega \cdot m$	58.8	65.9	72.9	79.5	86.0	92.2	98.3	104.1	09.9	115.5
$a \cdot 10^6$, m^2/s	22.0	22.0	22.0	22.0	21.5	20.7	20.0	19.0	18.0	—
λ, W/(m·K)	55.8	57.2	58.6	60.0	61.4	62.8	64.2	65.6	67.0	—

THE ELASTIC CHARACTERISTICS OF METALS OF THE VANADIUM SUBGROUP

The elastic constants of vanadium [220], niobium [221] and tantalum [222] have been measured for the temperature interval starting at the helium temperature and up to the melting point (see Fig. 192 and 193). The unusual behaviour of the trigonal shear constant c_{44}, shown schematically in Fig. 193, is specific to all of these metals. Similar anomalous behaviour of the elastic constant c_{44} was observed for the 3d-, 4d- and 5d-transition metals with the fcc structure (Pd, Pt) and the bcc structure (Va, Nb, Ta). This similarity in $c_{44}(T)$ arises from the common features of the Fermi surface of these metals: the Fermi level lies near the maximum of the density of states, and the Fermi surface is partly made up of open and closed sheets. The anomalous behaviour of the transverse acoustic phonons in the phonon spectrum of niobium is related to the anomaly in the dependence of $c_{44}(T)$ [221]. According to the data of various authors, the elastic moduli of polycrystalline vanadium, tantalum and niobium, at 300 K, differ considerably. For example, for tantalum the values of E vary from 1.75 to $1.87 \cdot 10^{11}$ Pa; $G = (0.685 \div 0.774) \cdot 10^{11}$ Pa; for niobium: $E = (0.85 \div 1.57) \cdot 10^{11}$ Pa; $G = (0.31 \div 0.87) \cdot 10^{11}$ Pa.

The values and the temperature dependence of moduli are influenced by the structure and history of the samples, and their purity.

Study [223] estimated the influence of the structure and interstitual impurities (oxygen, nitrogen, carbon) on the magnitude of the elastic moduli for a broad temperature interval. The data obtained for niobium by different methods are given in Table 37. The unusual increase in moduli with temperature up to $T \approx 1500$ K may be caused by the presence of interstitual impurities and by considerable anisotropy in the elastic properties of niobium.

The elastic moduli and Poisson's coefficient, from room temperature to 2070 K, are given below. The data were obtained with samples of technically pure tantalum produced by vacuum-arc melting [223].

T, K	293	370	470	570	670	770	870	970	1070	1170
$E \cdot 10^{-10}$, Pa	17.5	17.3	17.1	16.9	16.7	16.5	16.4	16.2	16.1	16.0
$G \cdot 10^{-10}$, Pa	7.05	6.95	6.87	6.85	6.75	6.65	6.55	6.55	6.45	6.45
μ	0.24	0.24	0.25	0.24	0.25	0.24	0.25	0.23	0.24	0.24

T, K	1270	1370	1470	1570	1670	1770	1870	1970	2070
$E \cdot 10^{-10}$, Pa	15.8	15.7	15.6	15.5	15.3	15.2	15.0	14.9	14.8
$G \cdot 10^{-10}$, Pa	6.35	6.35	6.27	6.27	6.18	6.18	6.08	6.08	5.48
μ	0.24	0.23	0.24	0.24	0.24	0.23	0.23	0.23	0.23

Table 37 Young's modulus ($E \cdot 10^{-10}$, Pa), shear modulus ($G \cdot 10^{-10}$, Pa) and Poisson's coefficient (μ) of polycrystalline niobium [223]

T, K	E^*/E^{**}	G^*	μ^*	T, K	E^*/E^{**}	G^*	μ^*
293	10.3/11.0	3.82	0.35	1470	10.3/10.7	3.92	0.31
370	10.3/10.9	3.82	0.35	1570	10.2/10.7	3.82	0.33
470	10.3/10.9	3.92	0.31	1670	10.0/10.7	3.82	0.31
570	10.3/10.9	3.92	0.31	1770	9.8/10.6	—	—
670	10.3/10.8	3.92	0.31	1870	9.6/10.5	—	—
770	10.3/10.8	3.92	0.31	1970	—/10.4	—	—
870	10.4/10.8	3.92	0.33	2070	—/10.2	—	—
970	10.4/10.9	3.92	0.33	2170	—/10.1	—	—
1070	10.5/10.9	3.92	0.34	2270	—/9.9	—	—
1170	10.5/11.0	3.92	0.34	2370	—/9.8	—	—
1270	10.5/10.8	3.92	0.34	2470	—/9.6	—	—
1370	10.4/10.8	3.92	0.35				

*For compact niobium.
**For cast niobium.

At T $\approx$ 650 K there is a visible change in the slope of $E(T)$. Similar behaviour is observed for tantalum. Measured using a wire sample, the Young's modulus E at $T <$ 600 K decreases considerably with increases in temperature; from 700 to 2100 K the temperature dependence of E (in Pa) is described by the following equation [224].

$$E = [(1.69 - 8.22 \cdot 10^{-5} T - 1.66 \cdot 10^{-8} T^2) \pm 0 \ 05] \cdot 10^{11}$$

The discontinuity in the curve of E at 700 K can be explained by a relaxation of the modulus, caused by dissolved oxygen. This relaxation was observed earlier for niobium and tantalum, where oxygen was added for that purpose. The investigated sample of tantalum contained the following impurities in ppm: C $-$ 10, O_2 $-$ 3, H_2 $-$ 1, N_2 $-$ 3.

The shear modulus of tantalum, in the temperature region from 300 to 750 K, can be described by the following equation [224].

$$G = [(0.774 - 1.73 \cdot 10^{-4} T) \pm 0.016] \ 10^{11}$$

The temperature dependences of the E and G moduli (in Pa) can be expressed for vanadium as

$$E = [(1.28 - 9.61 \cdot 10^{-5} T) \pm 0.040] \ 10^{11} \quad \text{at} \quad T = (300 - 1850) \ K$$
$$G = [(0.488 - 8.43 \cdot 10^{-5} T) \pm 0.011] \ 10^{11} \quad \text{at} \quad T = (300 - 1000) \ K$$

and for niobium

$$E = [(1.00 + 9.18 \cdot 10^{-5} T - 4.11 \cdot 10^{-8} T^2) \pm 0.028] \ 10^{11}$$
$$\text{at} \quad T = (300 - 1950) \ K$$
$$G = [(0.312 + 9.90 \cdot 10^{-6} T) \pm 0.001] \ 10^{11} \quad \text{at} \quad T = (300 - 750) \ K$$

In the above expressions the thermal expansion and the changes in density with temperature are taken into account.

§6 CHROMIUM, MOLYBDENUM, TUNGSTEN

CHROMIUM

At standard pressure, chromium has a bcc structure with the lattice parameter a = 0.28845 nm at 293 K [39, 74]. Near 312 K, an antiferromagnetic ordering takes place. The Neel temperature depends appreciably on the thermal history and purity of the samples. The anomaly in the magnetic and other properties at 123 K is due to a change in the polarization of the spin density waves.

The Fermi level of chromium lies near the minimum of the density of the electronic states (see Fig. 81). The Fermi surface consists of a large closed surface of holes at the H point in the third zone, small closed surfaces of holes at the N point in the third zone, a large closed electronic surface at the Γ point in the fourth zone, and small electronic pockets or lenses in the fifth zone, positioned along the ΓH line [41].

The temperature dependences of the elastic constants c_{11}, c_{44}, c' and Young's modulus, shown in Fig. 203, have well pronounced anomalies in the magnetic phase transition regions at $T_N = 311$ K and $T_S = 123$ K. Broken and broken-dotted lines show the data of earlier investigations. According to [229], the difference between the temperature dependences of c_{ij} is caused by different purities of the samples. In [229] the chromium used was 99.99% pure. From 123 to 311 K, in the antiferromagnetic phase with the transversely polarized spin density waves, the decrease in hardness can be suppressed by imposing a sufficiently strong magnetic field. This reveals the cause of this anomaly: the motion of the domain-boundary walls.

The values of the elastic constants at $T = 300$ K, given in [135], are: $E = 2.40 \cdot 10^{11}$ Pa; $G = 0.90 \cdot 10^{11}$ Pa; $K = 1.965 \cdot 10^{11}$ Pa.

The temperature dependence of the electrical resistivity of chromium is shown in Fig. 204 [231–237]. Its shape differs considerably from the shape that is typical for metals of the III and IV subgroups; $\partial^2\rho/\partial T^2 > 0$. Near the Neel point, the electrical resistivity and its derivative have an anomaly, shown in detail in Fig. 204. Near the Curie point, and below it, the magnetic and thermomechanical history of polycrystalline samples of chromium substantially influence many properties.

The electrical resistivity of paramagnetic single crystals, containing 26 ppm

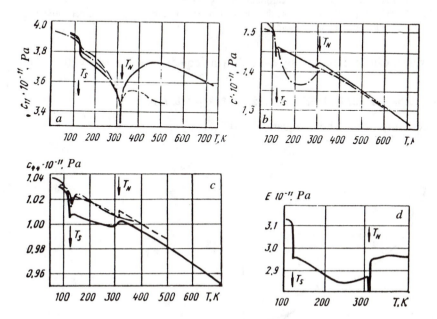

Figure 203 The elastic constants of chromium vs. temperature: a) c_{11} [229]; b) $c' = (c_{11} - c_{12})/2$ [229]; c) c_{44} [229]; d) E [230].

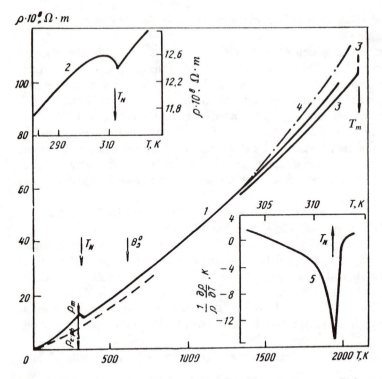

Figure 204 The electrical resistivity of chromium (ρ) vs. temperature: *1*) the average data of [140, 231, 232, 235, 237]; *2*) [232]; *3*) [237]; *4*) the calculation using equation (11); *5*) the derivative with respect to temperature near the Curie point.

of impurities, can be described by the following equation, valid from 420 to 1300 K [235]:

$$\rho(T) = 1.4087 + 2.9373 \cdot 10^{-2}T + 8.1752 \cdot 10^{-6}T^2 + 1.827 \cdot 10^{-9}T^3$$

The uncertainty of the above equation is 0.2%.

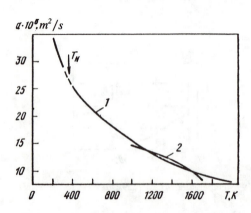

Figure 205 The diffusivity of chromium (*a*) vs. temperature: *1*) [39]; *2*) [140].

Using the model of electron scattering by disordered spins, the electrical resistivity in the paramagnetic region can be expressed as the following sum $\rho = \rho_m + \rho_{e-ph} + \rho_i$. The first term is the magnetic contribution: $\rho_m < \rho$, $\rho_m = 1.4$ $\mu\Omega \cdot cm$, which is of the same order as the value given in study [15] $\rho_m \approx 3 \div 4$ $\mu\Omega \cdot cm$. For paramagnetic chromium the small magnitude of the magnetic contribution makes it insignificant against the experimental uncertainty at moderate and high temperatures. Above 1300 K, there is a deviation from the above equation which was explained in [231] by a change in the position of the Fermi level.

The thermal conductivity of chromium is given in Fig. 206 and at the end of this section, its uncertainty is $\approx 10\%$. At low and moderate temperatures the Lorenz number differs appreciably from the standard value of L_0 (see Fig. 207). Above 200 K, the Lorenz number L is greater than L_0, above 1000 K $L \approx L_0$. According to several different investigators, the anomaly in the thermal conductivity near the Neel point is poorly reproducible and depends on the thermal history of the samples (see Fig. 206). Study [231] showed that Mott's model of s-d scattering fairly well describes the behaviour of the electrical resistivity of chromium and also explains the values for its thermal conductivity and Lorenz number.

The temperature dependence of the thermoelectric power of chromium is given in Fig. 208. It has an extreme point and an inversion point at high temperatures.

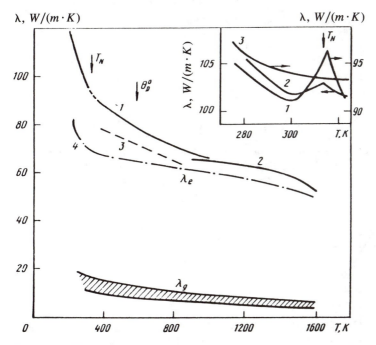

Figure 206 The thermal conductivity of chromium (λ) vs. temperature: *1*) [48]; *2*) [140]; *3*) λ_e^g; 4 — λ_e^L, λ_g—the estimation of [140]; in the inset—the thermal conductivity near the Curie point: 1—measurements before annealing [232]; 2—the same after annealing; 3—measurements of Moore et al. [233] who specifically investigated the shape of the $\lambda(T)$ anomaly near T_N.

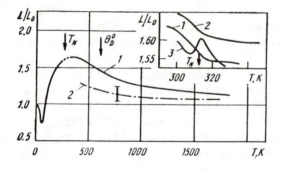

Figure 207 The Lorenz function of chromium (L) vs. temperature: 1) [231]; 2) the same, taking into account the lattice contribution; in the inset: 1 and 2) [233]; 3) [232].

This can be explained by the changes in the electronic spectrum and the position of the Fermi level [231]. At room temperature, the Hall coefficient of chromium is $3.63 \cdot 10^{-10}$ m^3/C [82, 100].

The kinetic properties of chromium are presented below [39, 48, 140, 231–235, 237].

T, K	100	200	300	400	500	600	700	800
$\rho \cdot 10^8$, $\Omega \cdot$ m	3	6	12	14	19	23	28	32
$a \cdot 10^6$, m/s	11.5	40.0	29.0	25.4	23.3	22.5	19.8	18.2
λ, W/(m·K)	—	120	96	88	83	78	74	71

T, K	900	1000	1200	1400	1600	1800	2000
$\rho \cdot 10^8$, $\Omega \cdot$ m	37	42	52	63	74	87	102
$a \cdot 10^6$, m/s	16.7	15.3	13.0	11.2	9.72	8.55	7.62
λ, W/(m·K)	68	65	63	60	53	—	—

MOLYBDENUM

At standard pressure, molybdenum has a bcc structure with the lattice parameter $a = 0.31467$ nm at 293 K. The Fermi level of molybdenum, as in chromium,

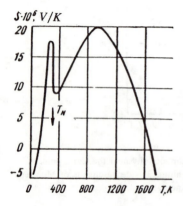

Figure 208 The thermoelectric power of chromium (S) vs. temperature [55].

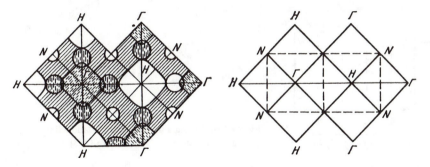

Figure 209 The Fermi surface of molybdenum [41].

lies near the minimum of the density of the electronic states (see Fig. 81). The Fermi surface of molybdenum is shown in Fig. 209.

The elastic moduli of molybdenum, obtained by different methods, have the following values: $E = (3.10 \div 3.30) \cdot 10^{11}$ Pa; $G = (1.19 \div 1.24) \cdot 10^{11}$ Pa. They were measured in [223] over the 300 to 2570 K range using samples cast and prepared through powder metallurgy. Table 38 summarizes the results.

The dependences of $E(T)$ and $G(T)$ were determined in [238] for the range from 300 to 1270 K:

$$E = [(3.12 - 4.18 \times 10^{-4}\, T) \pm 0.63]\, 10^{11} \text{Pa}$$
$$G = [(1.27 - 2.23 \cdot 10^{-4}\, T) \pm 0.033]\, 10^{11}\, \text{Pa}$$

Figure 210 shows the elastic moduli versus temperature for single crystals of molybdenum [223]. Near room temperatures $(\partial E/\partial T)_{\text{single}} < (\partial E/\partial T)_{\text{poly}}$.

Study [239] gave the following values for the elastic constants and the bulk

Table 38 Young's modulus ($E \cdot 10^{-10}$, Pa), shear modulus ($G \cdot 10^{-10}$, Pa) and Poisson's coefficient (μ) of polycrystalline molybdenum [223]

T K	Sintered Molybdenum			Cast Molybdenum (arc welding)			T, K	Sintered Molybdenum			Cast Molybdenum (arc welding)		
	E	G	μ	E	G	μ		E	G	μ	E	G	μ
293	3.22	1.21	0.32	3.22	1.19	0.335	1470	2.54	0.96	0.32	2.68	0.99	0.356
370	3.15	1.20	0.31	3.18	1.18	0.350	1570	2.48	0.94	0.31	2.63	0.96	0.366
470	3.11	1.18	0.32	3.14	1.16	0.355	1670	2.37	0.92	0.28	2.55	0.94	0.354
570	3.06	1.16	0.32	3.10	1.14	0.360	1770	2.22	—	—	2.47	0.91	0.354
670	3.00	1.14	0.31	3.06	1.12	0.366	1870	2.08	—	—	2.38	0.89	0.334
770	2.94	1.12	0.31	3.01	1.11	0.360	1970	1.93	—	—	2.31	0.87	0.341
870	2.89	1.10	0.31	2.97	1.10	0.352	2070	1.77	—	—	2.22	0.82	0.351
970	2.82	1.07	0.32	2.92	1.08	0.355	2170	1.63	—	—	2.16	0.79	0.358
1070	2.76	1.05	0.31	2.88	1.06	0.360	2270	—	—	—	2.08	0.75	0.375
1170	2.70	1.03	0.31	2.84	1.05	0.354	2370	—	—	—	2.00	—	—
1270	2.65	1.03	0.31	2.80	1.03	0.360	2470	—	—	—	1.92	—	—
1370	2.58	0.98	0.32	2.75	1.01	0.364	2570	—	—	—	1.84	—	—

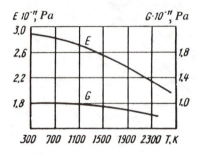

Figure 210 Young's (E) and shear (G) moduli of single crystal molybdenum vs. temperature [223].

modulus of molybdenum for $T = 300$ K: $c_{11} = 4.617 \cdot 10^{11}$ Pa; $c_{12} = 1.647 \cdot 10^{11}$ Pa; $c_{44} = 1.087 \cdot 10^{11}$ Pa; $c' = 2.970 \cdot 10^{11}$ Pa; $K = 2.637 \cdot 10^{11}$ Pa.

Study [240] compiled and assessed the most reliable literature data on the electrical resistivity of molybdenum. The results are shown in Fig. 211 and summarized at the end of this section. The influence of impurities on resistivity at 300 K was expressed in [240] as $\rho(c) = 5.553 + 3.31c$ $\mu\Omega \cdot$ cm where c is the total impurity content in % (by mass).

In the liquid state, for $T_m < T < 7000$ K, the electrical resistivity (in $\mu\Omega \cdot$ m) is described by the following expression [210]

$$\rho = 0.9710 + 7.9913 \cdot 10^{-5} (T - T_m) + 8.0148 \cdot 10^{-9} (T - T_m)^2$$

where $T_m = 2890$ K.

The main distinguishing feature of the temperature dependence of the electrical resistivity of molybdenum is its positive curvature ($\partial^2\rho/\partial T^2 > 0$). Study [11] calculated this dependence using Mott's theory of s-d scattering and taking into account the parameters of the Fermi surface and thermal expansion. The results fairly well describe the dependence of $\rho(T)$ (see Fig. 211). Thermal expansion is the main factor leading to $\partial^2\rho/\partial T^2 > 0$.

Above 200 K, the diffusivity of molybdenum at temperatures decreases [39, 140, 228]. The uncertainty in the data, given in Fig. 212 and at the end of this section, is about 5–7%.

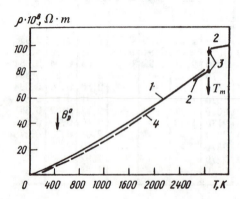

Figure 211 The electrical resistivity of molybdenum (ρ) vs. temperature: *1*) the average data of [240]; *2*) [210]; *3*) [241]; *4*) the calculation based on the s-d model [11].

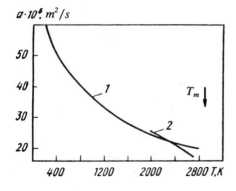

$a \cdot 10^6$, m^2/s

Figure 212 The diffusivity of molybdenum (a) vs. temperature: _1_) [39]; _2_) [228].

At moderate and high temperatures, the thermal conductivity of molybdenum also has a negative temperature coefficient. The data of [226] given in Fig. 213 correspond to the most reliable investigations, which used samples of 99.9–99.99% purity; the uncertainty is ≈ 8–10%. Calculation [11] showed that the temperature dependence of the thermal conductivity can be described quite well by Mott's model of the s-d scattering. From the relationship between λ_e^g and λ_e^L it is clear that the Lorenz number of molybdenum is less than L_0 below 200 K. It however, reaches a maximum at moderate temperatures and then approaches the value of L_0. The deviation from the value of L_0 at low temperatures is caused by the inelastic scattering of electrons. Taking into account the lattice contribution to thermal conductivity, it can be estimated that the value of L exceeds that of L_0 by 10–15% at moderate temperatures. This is caused by band effects [11].

The temperature dependence of the thermoelectric power of molybdenum is of an extreme nature [55]. Calculation [11], carried out with the parameters given below and using equation (28), produces the same qualitative results (Fig. 214).

At room temperature, the Hall coefficient of molybdenum is $1.91 \cdot 10^{-10}$ $m^3/$C [82].

The kinetic properties of molybdenum, listed below, take into account the thermal expansion [39, 48, 140, 210, 240–242]

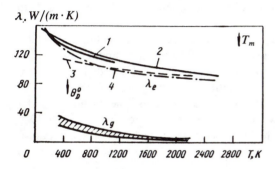

λ, $W/(m \cdot K)$

Figure 213 The thermal conductivity of molybdenum (λ) vs. temperature: _1_) [48]; _2_) [242]; _3_) λ_e^g; _4_) λ_e^L; λ_g is the calculation of [140].

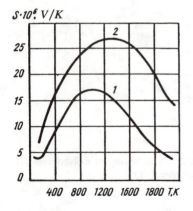

Figure 214 The thermoelectric power (*S*) of molybdenum vs. temperature: *1*) [55]; *2*) the calculation from the *s-d* model [11].

T, K		200	300	400	500	600	700
$\rho \cdot 10^8$, $\Omega \cdot m$		—	5.55	8.00	10.4	13.0	15.6
$a \cdot 10^6$, m^2/s		60.0	54.8	50.6	47.6	45.0	43.0
λ, W/(m·K)		148.4	145.4	140.5	135.8	131.7	127.7

T, K		800	900	1000	1200	1400
$\rho \cdot 10^8$, $\Omega \cdot m$		18.2	20.9	23.6	29.2	34.9
$a \cdot 10^6$, m^2/s		41.0	38.8	37.2	33.6	30.8
λ, W/(m·K)		124.1	120.8	118.0	112.5	108.4

T, K	1600	1800	2000	2200	2400	2600	2800
$\rho \cdot 10^8$, $\Omega \cdot m$	40.8	47.0	53.3	59.8	66.5	73.5	80.5
$a \cdot 10^6$, m^2/s	28.5	26.5	24.8	23.2	21.0	18.7	—
λ, W/(m·K)	104.7	101.7	99.2	96.9	95.0	93.3	—

TUNGSTEN

Tungsten is the most refractory metal; it has a bcc structure with the lattice parameter $a = 0.31649$ nm at 298 K. The electronic structure of tungsten has not been sufficiently studied [41]. Its Fermi level lies near the minimum of the curve of density of the electronic states (see Fig. 93). The Fermi surface of tungsten differs from those of chromium and molybdenum by the smaller size of all of the sheets and by the presence of small or disappearing electronic lenses in the fifth zone, along the ΓH direction.

Different investigations obtained the following values for moduli at $T = 300$ K: $E = (3.4 \div 5.00) \cdot 10^{11}$ Pa $G = (1.4 \div 1.8) \cdot 10^{11}$ Pa, depending on the method of preparation and the thermal treatment of the samples. Nevertheless, the appearance of the temperature dependence is the same, regardless of the method of sample preparation (Table 39). As temperature increases, the value of E decreases, and $\partial E/\partial T$ increases. However, at high temperatures (greater than 2000 K), the values of E for tungsten are higher than those of steel at room temperature.

Table 39 Young's ($E \cdot 10^{-10}$, Pa) and shear ($G \cdot 10^{-10}$, Pa) moduli and Poisson's coefficient (μ) of polycrystalline tungsten [223]

T, K	Sintered Tungsten			Cast Tungsten			T, K	Sintered Tungsten			Cast Tungsten		
	E	G	μ	E	G	μ		E	G	μ	E	G	μ
273	3.90	1.45	0.34	4.04	1.60	0.264	1670	3.25	1.20	0.36	3.50	1.35	0.293
370	3.85	1.43	0.34	4.00	1.59	0.260	1770	3.14	1.17	0.34	3.44	1.33	0.290
470	3.80	1.41	0.34	3.96	1.57	0.261	1870	3.06	—	—	3.35	1.30	0.285
570	3.76	1.39	0.35	3.93	1.55	0.269	1970	2.95	—	—	3.28	1.27	0.288
670	3.72	1.37	0.35	3.89	1.54	0.265	2070	2.84	—	—	3.21	1.24	0.291
770	3.68	1.35	0.35	3.85	1.52	0.269	2170	2.74	—	—	3.14	1.22	0.291
870	3.63	1.33	0.35	3.81	1.50	0.270	2270	2.65	—	—	3.06	1.19	0.290
970	3.49	1.32	0.35	3.77	1.48	0.274	2370	2.55	—	—	2.98	1.16	0.290
070	3.54	1.31	0.34	3.74	1.46	0.280	2470	2.45	—	—	2.90	1.12	0.291
1170	3.50	1.29	0.35	3.70	1.45	0.280	2570	—	—	—	2.83	1.10	0.290
1270	3.45	1.27	0.35	3.67	1.43	0.280	2670	—	—	—	2.75	1.07	0.290
1370	3.41	1.25	0.35	3.63	1.41	0.285	2770	—	—	—	2.68	1.04	0.290
1470	3.36	1.23	0.36	3.60	1.39	0.290	2870	—	—	—	2.60	1.01	0.292
1570	3.31	1.22	0.36	3.55	1.37	0.293							

The elastic moduli of cast tungsten are higher than those of sintered tungsten over the whole temperature range.

The temperature dependence of the elastic moduli of single crystal tungsten differs somewhat from that of polycrystalline samples: $(\partial E / \partial T)_{single} < (\partial E / \partial T)_{poly}$, as the temperature increases, the moduli change more smoothly (Fig. 215).

The electrical resistivity was investigated using mainly samples of 99.99% purity. The resistivity has been reasonably well determined, except near the melting point, where it requires some verification [82, 227, 243, 244].

According to [243], tungsten can be used as a standard substance in the investigations of electrical and thermal conductivities. The recommended values of electrical resistivity are given at the end of this section. Study [210] measured the electrical resistivity of liquid tungsten for temperatures below 7500 K, the results were described by the following equation (in $\mu\Omega \cdot m$): $\rho = 1.350 - 1.885 \cdot 10^{-5}$ $(T - T_m) + 4.42 \cdot 10^{-8} (T - T_m)^2$ where $T_m = 3680$ K.

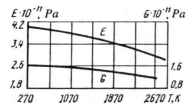

Figure 215 The variation of Young's (E) and shear (G) moduli with temperature for single crystal tungsten [223].

Study [11] calculated using Mott's *s-d* model and taking into account thermal expansion (changing the Debye temperature), electrical resistivity as a function of temperature for the Fermi surface with four zones. The results of the calculations agree quite well with the experimental data. The main factor accounting for the increase in $\partial\rho/\partial T$ is thermal expansion (Fig. 216).

The diffusivity of tungsten, given in Fig. 217 and at the end of this section, was determined with an uncertainty of 5% for the interval from 300 to 1500 K, 8% below 300 K and from 1500 to 3000 K, 14% above 300 K. Above 2500 K, the preference was given to the data of [228] (curve 2 in Fig. 217), however, additional investigations are required for this temperature range.

Above 100 K, the thermal conductivity of tungsten is also a decreasing function of temperature [48, 243] (Fig. 218). At moderate temperatures, the Lorenz number is even greater than the standard value of L_0 (Fig. 219). Calculations [11, 244] gave a qualitatively similar temperature dependence of λ_e^L.

The temperature dependence of the thermoelectric power of tungsten, as was the case for chromium and molybdenum, is of an extreme nature (Fig. 220). The calculations of [11] gave the same shape of $S(T)$, but substantially overestimated the values at $T = 1500$ K. The Hall coefficient of tungsten changes monotonically from $1.06 \cdot 10^{-10}$ m^3/C at 173 K to $1.11 \cdot 10^{-10}$ m^3/C at 273 K and $1.56 \cdot 10^{-10}$ m^3/C at 873 K [100].

The kinetic properties of tungsten are summarized below [39, 210, 228, 243].

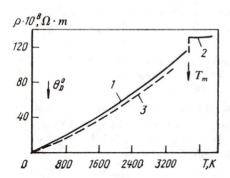

Figure 216 The electrical resistivity of tungsten (ρ) vs. temperature: *1*) [243]; *2*) [210]; *3*) the calculation of [11] using Mott's *s-d* model.

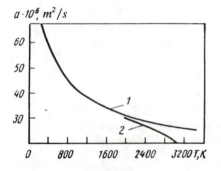

Figure 217 The diffusivity of tungsten (*a*) vs. temperature: *1*) [39]; *2*) [228].

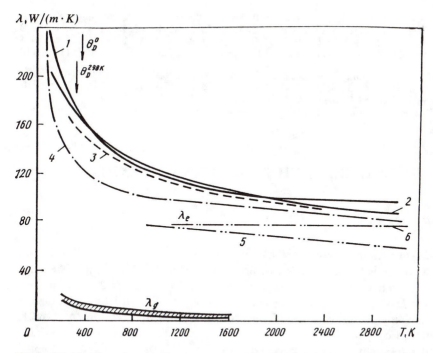

Figure 218 The thermal conductivity of tungsten (λ) as a function of temperature: *1*) [48]; *2*) the average data of [243]; *3*) λ_e^g; *4*) λ_e^L; *5*) λ_g, calculated in [244]; *6*) the same as 5, but taking into account the temperature dependence of θ_D.

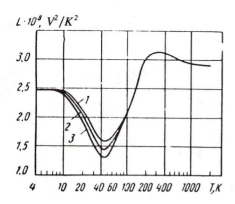

Figure 219 The variation of the Lorenz function (L) of tungsten with temperature: *1–3*) the data for samples of different purity: $r = 20, 50, 100$, respectively.

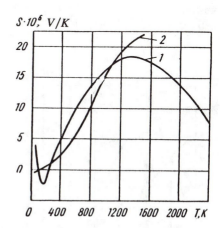

Figure 220 The thermoelectric power of tungsten (S) vs. temperature [55]: *1*) [151]; *2*) the calculation of [11].

T, K	200	300	400	500	600	700	800	900	1000	1200	1400
$\rho \cdot 10^8$, $\Omega \cdot$m	3.19	5.45	7.84	10.3	12.9	15.6	18.5	21.4	24.4	30.5	36.9
$a \cdot 10^6$, m²/s	—	64.0	59.6	54.9	51.0	47.9	45.2	43.2	41.4	38.6	36.4
λ, W/(m·K)	—	172	158	147	139	—	127	—	120	114	110

T, K	1600	1800	2000	2200	2400	2600	2800	3000	3200	3400	3600
$\rho \cdot 10^8$, $\Omega \cdot$m	43.5	50.2	57.1	64.0	71.0	78.1	85.3	92.6	100	107	115
$a \cdot 10^6$, m²/s	34.5	32.6	30.4	28.6	27.0	25.4	23.5	21.2	—	—	—
λ, W/(m·K)	107	105	102	—	99	—	97	97	—	—	—

§7 MANGANESE, TECHNETIUM AND RHENIUM

MANGANESE

At standard pressure, manganese has four crystalline modifications. Below 1100 ± 20 K, α-Mn is stable and has a complex cubic structure consisting of 58 atoms in the unit cell with the lattice parameter $a = 0.89136$ nm at 293 K. Between 1100 and 1360 ± 10 K, β-Mn is stable, it also has a cubic structure with the unit cell containing 20 atoms and the lattice parameter $a = 0.33144$ nm. At 1368 K, an hcp phase, γ-Mn, is formed with $z = 2$ and the lattice parameter being $a = 0.38623$ nm. Finally, between 1410 ± 5 K and the melting point at 1517 ± 5 K, the δ-Mn modification is stable and has a bcc structure with the lattice parameter $a = 0.3805$ nm at 1413 K. At 95 K α-Mn transforms into the antiferromagnetic state.

Judging by the data on the low-temperature electronic heat capacity (see Fig. 81), the density of the electronic states near the Fermi level for α-Mn is the highest among the transition metals. The Fermi surface of manganese has not been investigated either experimentally or theoretically because its crystalline and magnetic structures are complex.

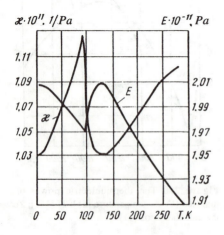

Figure 221 Young's modulus (E) and the volume compressibility ($\varkappa$) of manganese as a function of temperature [245].

Figure 221 shows Young's modulus and the volume compressibility of 99.9% pure polycrystalline manganese for the temperature range from 4 to 300 K. As temperature decreases, the E modulus increases. At ≈ 125 K, the value of the E modulus decreases, and after reaching a minimum at $T_N = 96$ K, it increases again. The minimum on the curve of $E(T)$ corresponds to the peak of $\approx 6\%$ in the volume compressibility curve. Table 40 lists Young's modulus of annealed manganese from room temperature to 1073 K.

Figure 222 shows the electrical resistivity of manganese. The available information is quite contradictory, especially for the 'old' results [246]. According to [246–248], the resistivity decreases during structural transformations and has a minimum value at high temperatures, for liquid manganese. A relatively recent investigation [247] gave for α-Mn, at room temperature, the value of 154 ± 6 $\mu\Omega \cdot$ cm. Studies [140] and [15] found similar values, the highest among the transition metals. The temperature coefficient of the electrical resistivity decreases sharply in the paramagnetic region. The low accuracy of the given values of the kinetic coefficients is due to the low purity of the samples ($r \approx 5$–10) [247, 15, 140].

At $T > T_N$ the main contribution to the electrical resistivity of α-Mn, $\rho = \rho_i + \rho_{e-ph} + \rho_m$, comes from scattering by the disordered spins $\rho_m = D$; $\rho = \rho_0 + AT + D$ (see Fig. 222).

The direct measurements of the diffusivity of manganese were performed only in [140] using polycrystals with $r = 5$. The uncertainty in the measurements was $\approx 5\%$. However, considering the relatively low purity of the samples, the data require refinement (Fig. 223).

The information on the thermal conductivity of manganese is shown in Fig. 224. The data require verification because the samples were of low purity. The positive temperature coefficient is due to the increase in $\lambda_e^L = L_0(A + DT^{-1})^{-1}$. It should be mentioned that, at room temperature, the thermal conductivity of manganese is the smallest among transition metals. According to [140], the lattice component dominates almost completely in the overall conductivity. This is because the electronic component is small, which is unusual for metals. These data indicate indirectly that, at room temperature, for metals of the manganese subgroup,

Table 40 The elastic characteristics of metals of the manganese subgroup

T, K	Manganese [82]	Technetium [249]			
	$E \cdot 10^{-11}$, Pa	$E \cdot 10^{-11}$, Pa	$G \cdot 10^{-11}$, Pa	$K \cdot 10^{-11}$, Pa	μ
4.2	—	3.41	1.25	3.06	0.320
78	—	3.28	1.24	2.96	0.316
298	1.98	3.22	1.23	2.81	0.309
473	1.86	—	—	—	—
673	1.69	—	—	—	—
873	1.43	—	—	—	—
1073	1.22	—	—	—	—

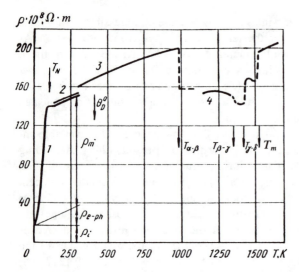

Figure 222 The electrical resistivity of manganese (ρ) vs. temperature: *1*) [246]; *2*) [247]; *3*) [140]; *4*) [248].

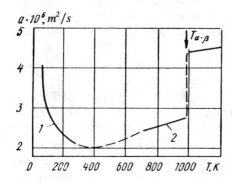

Figure 223 The diffusivity of manganese (*a*) vs. temperature: *1*) [39]; *2*) [140].

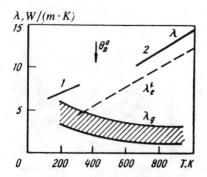

Figure 224 The thermal conductivity of manganese (λ) vs. temperature: *1*) [48]; *2*) [140].

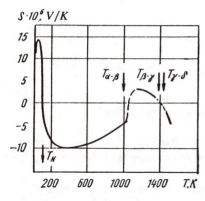

Figure 225 The absolute value of the thermoelectric power of manganese (S) as a function of temperature [55].

the lattice component amounts to a few W/m. This can be confirmed by using the estimates from equations (21) and (22).

The thermoelectric power of manganese was investigated in [55]. The absolute value of the thermoelectric power is small, and its temperature dependence has several inversion points (Fig. 225). At the room temperature, the Hall coefficient of manganese is equal to $0.844 \cdot 10^{-10}$ m^3/C [82].

TECHNETIUM

Technetium does not have any stable isotopes. The half-life of the longest-lived ^{99}Tc is $2.1 \cdot 10^5$ years. Technetium emits β-particles with an energy of 0.29 MeV. At standard pressure, technetium has an hcp structure with the lattice parameters $a = 0.2735$ nm and $c = 0.4388$ nm at 293 K [74].

The Fermi level of technetium lies near the maximum of the density of the electronic states, although the value of $N(\varepsilon)$ is substantially smaller than that of

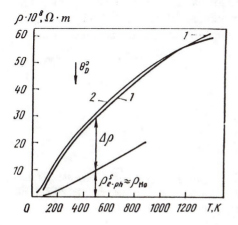

Figure 226 The electrical resistivity of technetium (ρ) vs. temperature: *1*) [251]; *2*) [140].

$a \cdot 10^6, m^2/s$

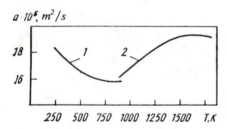

Figure 227 The diffusivity of technetium (a) vs. temperature: *1*) [39]; *2*) [250].

manganese (see Fig. 50). The Fermi surface of technetium has not been investigated. It is assumed to be similar to that of rhenium if the contribution from the effects of relativity is neglected [41].

The elastic characteristics of polycrystalline technetium are given in Table 40. The investigated specimen contained the following impurities, in parts per million: C—4, H_2—4, N_2—3, O_2—9.

The electrical resistivity of polycrystalline technetium of 99.9% purity was investigated in [250, 251] (in [250] the impurities amounted to $\approx 0.08\%$); the resistivity is given in Fig. 226 and at the end of this section. The error in the measurements was $\approx 2\%$. Nevertheless, the data require verification because the thermomechanical treatment and the anisotropy in resistivity affect the results. If the anisotropy in resistivity of technetium does not exceed that of rhenium, then the uncertainty in the resistivity may be estimated as 5% at moderate temperatures.

Since in manganese the contribution from the d-band is relatively small, this component can be isolated in technetium by comparing the resistivities of manganese and technetium (see Fig. 226). Only at high temperatures does the electron-phonon component of technetium, ρ_{e-ph}, computed from the resistivity of molybdenum, become comparable to the other contributions. At moderate temperatures, $\rho - \rho_{e-ph} > \rho_{e-ph}$. Other components $\Delta\rho = \rho - \rho_{e-ph}$ can be caused by the s-d band scattering or by scattering from the magnetic inhomogeneities of the paramagnon type [140].

The investigations of the diffusivity of polycrystalline technetium [39, 250] are in close agreement; the uncertainty can be estimated at 7–10% (Fig. 227).

The thermal conductivity of technetium has a wide minimum at moderate temperatures, which is typical for the transition metals [140, 252] (Fig. 228). The increase in the values of the thermal conductivity at high temperatures is due to

$\lambda, W/(m \cdot K)$

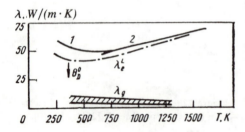

Figure 228 The thermal conductivity of technetium (λ) vs. temperature: *1*) [252]; *2*) [250]; λ_g) the calculation of [140].

the increase in its electronic component. However, at moderate temperatures, the lattice contribution may reach 10% of the total conductivity. Study [140] showed that estimating the lattice component using three different methods leads to approximately the same results. The first method is based on the standard W-F-L law above 1000 K; the second, takes into account the electron-phonon scattering; and, the third, uses the thermal conductivity of manganese at moderate temperatures, which is of the phonon nature. It is possible that, below 500 K, the value of L was greater than that of L_0 because of the band effects.

The thermoelectric power of technetium is a complex function with a minimum at low temperatures, a zero crossing, followed by a maximum point [55, 151, 250]. It is similar to the thermoelectric power of manganese and rhenium, despite several phase transformations in manganese.

The kinetic properties of technetium are summarized below [140, 250].

T, K	300	400	500	600	700	800
$\rho \cdot 10^8$, $\Omega \cdot$ m	20.4	25.7	30.8	35.5	39.8	43.9
$a \cdot 10^6$, m^2/s	17.8	17.1	16.6	16.2	16.0	16.0
λ, W/(m·K)	55	53	48	49	49	51

T, K	900	1000	1200	1400	1600	1800
$\rho \cdot 10^8$, $\Omega \cdot$ m	47.6	51.0	56.4	59.2	—	—
$a \cdot 10^6$, m^2/s	16.2	16.8	18.0	18.9	19.3	19.2
λ, W/(m·K)	52.5	55.0	59.0	63.0	67.0	—

RHENIUM

At standard pressure, rhenium has an hcp crystalline structure with the lattice parameters $a = 0.2760$ nm, $c = 0.4458$ nm at 293 K [74]. Rhenium is one of the most refractory metals ($T_m = 3463 \pm 20$ K).

The Fermi level of rhenium lies near the maximum of the density of electronic states (see Fig. 93). According to the calculations of Mattheiss [41], who took into account the relativistic band structure, the Fermi surface has five sheets: closed hole surfaces in the fifth, sixth and seventh zones and a closed electronic surface in the ninth zone. The electronic surface in the eighth zone is similar to a cylinder parallel to the $\langle 0001 \rangle$ axis.

The elastic constants of rhenium have been measured for the temperature range from 4 to 1100 K [38]. The temperature dependence of the values of c_{ij} are smooth (see the data at the end of this section). At $T = 298$ K polycrystalline rhenium has the following characteristics [82]: $E = 4.70 \cdot 10^{11}$ Pa; $G = 1.79 \cdot 10^{11}$ Pa; $K = 3.65 \cdot 10^{11}$ Pa.

The electrical resistivity of single crystal high purity rhenium at low temperatures has been extensively studied in [253]. The temperature dependence of the resistivity of rhenium, like that of technetium, has a negative slope at high temperatures (Fig. 229). The data for single crystals of rhenium, given below, were obtained by the present author for specimens with $r = 62$, cut parallel and per-

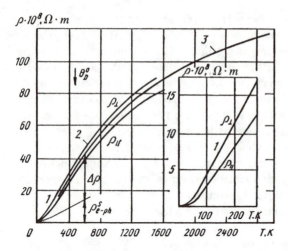

Figure 229 The temperature dependence of the electrical resistivity of rhenium in the direction of the hexagonal axis ($\rho_\parallel$) and perpendicular to it ($\rho_\perp$): *1*) [253]; *2*) [140]; *3*) the average data for polycrystalline rhenium [254, 255].

pendicular to the *c*-axis. The uncertainty in the data, given in Table 41 and Fig. 229, is 1.5% for the interval of 300–1600 K and about 3% at higher temperatures. Figure 229 also compares the electrical resistivities of rhenium and tungsten, the neighbor of rhenium in the periodic table. Both metals have similar Debye temperatures, ion masses and other characteristics; therefore, $\rho_{e-ph} \approx \rho_w$, in so far as

Table 41 The kinetic properties of rhenium [39, 48, 140, 253–255]

T. K	$\rho \cdot 10^8$, $\Omega \cdot m$			$a \cdot 10^6$, m^2/s		λ. W/(m·K)		
	poly	‖ *c*	⊥ *c*	‖ *c*	⊥ *c*	poly	‖ *c*	⊥ *c*
300	19.88	17.63	21.02	18.8*	16.6*	49	—	—
400	25.50	24.01	26.75	18.2*	16.4*	47	—	—
500	33.4	31.1	34.4	18.0*	16.3*	46	—	—
600	41.0	37.5	42.7	17.8*	16.2*	47*	—	—
700	48.2	44.2	50.2	17.8*	16.2*	48*	—	—
800	54.7	50.2	57.0	17.9	16.2*	50*	55.6	49.1
900	60.0	56.0	64.0	18.1	16.4	52*	—	—
1000	66.0	60.1	69.0	18.3	16.7	54*	58.2	53.1
1200	74.7	68.2	78.9	18.7	17.5	—	61.2	57.3
1400	81.2	75.7	84.0	19.1	18.0	—	64.1	60.4
1600	87.7	81.0	89.5	19.4	18.5	—	66.1	63.0
1800	94	—	—	19.6	18.7	—	68.4	65.3
2000	100	—	—	19.4	18.6	—	69.1	66.2
2200	104	—	—	18.8	18.2	—	68.3	66.1
2400	108	—	—	18.2	17.7	—	67.6	65.7
2600	111	—	—	17.5	17.1	—	66.3	64.8
2800	115	—	—	—	—	—	—	—
3000	118	—	—	—	—	—	—	—

*The data require improvement.

the resistivity of tungsten is determined mainly by the scattering of the s-type carriers. The remaining component of the resistivity is probably due to the high density of the electronic states of rhenium, caused by its d-band. It should be pointed out that, for temperatures above 2500 K, the electrical resistivities of tungsten and rhenium begin approaching each other.

Studies of the electron scattering processes in rhenium at moderate and low temperatures [253] showed that, in the region of $T \approx \theta_D$, the anisotropy of the electrical resistivity is determined by the anisotropy of its Fermi surface: $\rho_\perp/\rho_\parallel = S_\parallel/S_\perp = 1.21$. This is close to the ratio observed experimentally, $\rho_\perp/\rho_\parallel = 1.36$. Below the Debye temperature, the anisotropy in the resistivity is also determined by the geometry of the transition processes, and probably by the anisotropies of the phonon spectrum and matrix element of scattering.

The diffusivity of single crystals of rhenium in the direction of the hexagonal axis and perpendicular to it was obtained by the method of planar temperature waves [140] using crystals with $r = 62$, the uncertainty in these data is 3%.

The data given in Table 41 and Fig. 230 are close to those of [254] which were obtained for single crystals with axes that made an angle of 32° with the c-axis. The values of [39] for a polycrystalline sample determined at high temperatures seem to be too low.

Figure 231 and Table 41 show the temperature dependence of the thermal

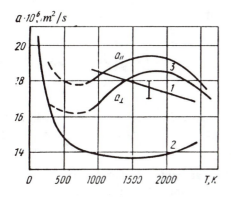

Figure 230 The diffusivity of rhenium in the direction of the hexagonal axis ($a_\parallel$) and perpendicular to it ($a_\perp$) vs. temperature: *1*) [254]; *2*) [39] for a polycrystalline sample; *3*) [140].

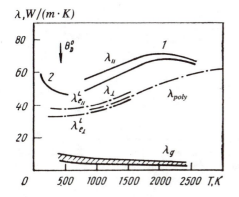

Figure 231 The thermal conductivity of rhenium in the direction of the hexagonal axis ($\lambda_\parallel$) and perpendicular to it ($\lambda_\perp$) vs. temperature: *1*) [140]; *2*) [48] for a polycrystalline sample; λ_g is the calculations of [140].

conductivity of single crystal rhenium along the c-axis and perpendicular to it, as well as the values of the electronic component in the same directions. The uncertainty of these data is 7%. The increase in the thermal conductivity of rhenium at high temperatures is caused by its electronic component. Below 2500 K, however, the difference between $\lambda_e^L = L_0 T/\rho$ and $\lambda_e^g = \lambda - \lambda_g$ is appreciable, and is probably due to band effects.

The temperature dependence of the thermoelectric power of rhenium is of an extreme nature with an inversion point at moderate temperatures (Fig. 232).

The Hall coefficient of rhenium changes monotonically from $0.4 \cdot 10^{-10}$ m^3/ C at 100 K to $1.6 \cdot 10^{-10}$ m^3/C at 300 K [100]. According to the data of [82], at 300 K the Hall coefficient of rhenium is equal to $3.15 \cdot 10^{-10}$ m^3/C.

The elastic constants of rhenium ($c_{ij} \cdot 10^{-11}$, Pa) are listed below [38].

T, K	4	173	298	373	473	573	673
c_{11}	6.446	6.311	6.182	6.103	6.002	5.903	5.882
c_{33}	7.170	7.001	6.835	6.748	6.641	6.541	6.450
c_{44}	1.685	1.648	1.606	1.582	1.549	1.514	1.479
c_{66}	1.838	1.783	1.714	1.676	1.626	1.579	1.533
c_{12}	2.770	2.745	2.753	2.752	2.750	2.745	2.756
c_{13}	1.959	2.026	2.078	2.098	2.118	2.138	2.154

T, K	773	873	923	1023	1073	1123
c_{11}	5.740	5.660	5.619	—	—	—
c_{33}	6.365	6.274	6.237	—	—	—
c_{44}	1.444	1.408	1.391	1.356	1.339	1.322
c_{66}	1.490	1.457	1.429	1.391	1.373	1.355
c_{12}	2.760	2.765	2.762	—	—	—
c_{13}	2.168	2.178	2.184	—	—	—

§8 RUTHENIUM, OSMIUM

RUTHENIUM

At standard pressure, up to the melting point, ruthenium has an hcp structure with the lattice parameters $a = 0.2705$ nm, $c = .4282$ nm at room temperatures [82].

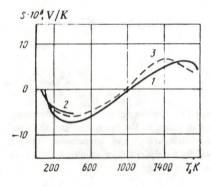

$S \cdot 10^6$, V/K

Figure 232 The temperature dependence of the thermoelectric power (S) of rhenium *1*) [250, 55], technetium *2*), and a combined dependence for the elements of the manganese subgroup *3*) [55].

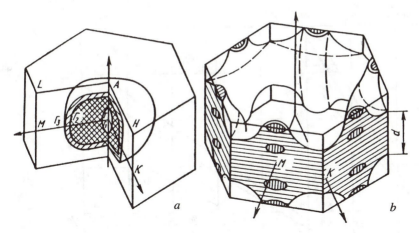

Figure 233 The Fermi surface of ruthenium [258]: *a*) the closed sheet of holes Γ_1, and the closed electronic sheets Γ_2 and Γ_3; *b*) the multiply-connected sheet of holes, the lenses of holes and closed ellipsoids of holes between *M* and *L*.

Its electronic structure has been studied rather well. According to the data of [41, 258], the Fermi surface of ruthenium consists of two closed electronic ellipsoids around the Γ point, three closed ellipsoids of holes around the *L*, *U* and Γ points, and an open surface of holes which is a combination of ellipsoids, compressed in the basal plane, centered around the *M* point, and interconnected by necks passing through the *L* and *K* points (Fig. 233). Judging by the data on the electronic heat capacity of ruthenium (see Fig. 87), the density of the electronic states near the Fermi energy is small, relative to the neighboring transition metals.

The elastic constants of hexagonal ruthenium have been measured from 4 to 923 K [38]. They are given below ($c_{ij} \cdot 10^{-11}$, Pa).

T, K . .	4	173	298	373	473	573	673	773	873	923
c_{11} . . .	5.763	5.699	5.626	5.580	5.516	5.448	5.378	5.306	5.234	5.204
c_{33} . . .	5.405	6.337	6.242	6.181	6.096	6.006	5.918	5.827	5.737	5.691
c_{44} . . .	1.891	1.853	1.806	1.778	1.735	1.691	1.647	1.609	1.557	1.534
c_{66} . . .	1.945	1.913	1.874	1.850	1.817	1.782	1.746	1.712	1.679	1.662
c_{12} . . .	1.872	1.873	1.878	1.882	1.883	1.882	1.885	1.880	1.880	1.880
c_{13} . . .	1.673	1.679	1.682	1.681	1.683	1.685	1.689	1.689	1.691	1.691

The temperature dependences of c_{ij} are smooth, without sharp irregularities. The elastic characteristics of polycrystalline ruthenium are given in Table 42.

The electrical resistivity of single crystal ruthenium below 300 K was studied in [253, 259], and above 290 K in [260]. In [259] the measurements were performed on samples with $r = 94$, 76.5 and 388 for the values of $\rho_\perp$, $\rho_\parallel$ and ρ_{poly}, respectively. In [253] $r = 880 \div 2000$ for different specimens and in [260] the measurements were performed on samples with $r = 600$. The uncertainty is $\rho \approx 1 \div 2\%$. Figure 234 and Table 43 show that dependences of $\rho(T)$ for $\rho_\perp$, $\rho_\parallel$ and

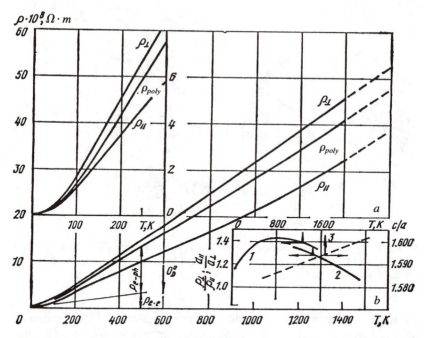

Figure 234 *a*) The electrical resistivity of ruthenium in the direction of the hexagonal axis ($\rho_\parallel$), perpendicular to it ($\rho_\perp$), and for a polycrystalline sample (ρ_{poly}) for temperatures below 300 K [253] and above 290 K [260]; *b*) the anisotropy in the resistivity: *1*) $\rho_\perp/\rho_\parallel$; *2*) $a_\parallel/a_\perp$; *3*) c/a.

ρ_{poly} are close to linear for the temperature range 300–1000 K; at higher temperatures, the values of $\partial\rho/\partial T$ start increasing. The anisotropy in resistivity increases with temperature, reaching a maximum value of $\rho_\perp/\rho_\parallel = 1.48$ at 700–900 K, then it decreases monotonically.

Study [140] indicated that the ratio of the lattice parameters (c/a) tends to increase, approaching the value of $c/a = 1.63$, the ideal for the hexagonal structures.

Table 42 The elastic characteristics of noble metals at 293 K [73]

Metal	$d \cdot 10^{-3}$, Kg/m³	v_l, m/s	v_t, m/s	$E \cdot 10^{-10}$, Pa	$G \cdot 10^{-10}$, Pa	μ
Pt	21.45	3988	1681	17.10	6.06	0.39
Pd	12.02	4372	1932	11.52	4.48	0.39
Ir	22.65	5285	3042	51.71	20.96	0.26
Rh	12.41	5604	3465	31.85	14.89	0.26
Os	22.61	5750	3192	55.85	22.25*	0.255*
Ru	12.45	6728	3668	41.37	16.30*	0.269**

*The data obtained from relationships for E and G.
**The value obtained from the equation $G = E/[2(1 + \mu)]$.

Table 43 The values of the electrical resistivity for iridium, ruthenium and osmium [140] ($\rho \cdot 10^8$, $\Omega \cdot$m)

T, K	Iridium	Ruthenium			Osmium		
		$\parallel c$	$\perp c$	poly	$\perp c$	$\parallel c$	poly
50	—	0.20	0.25	0.23	—	—	—
100	1.16	1.08	1.55	1.36	—	—	1.95
150	—	2.30	3.20	2.80	—	—	—
200	3.25	3.55	4.80	4.30	—	—	5.49
250	—	4.70	6.65	5.95	—	—	—
273	4.77	5.26	7.43	6.69	9.35	5.75	8.14
300	5.33	5.92	8.35	7.55	12.0	7.8	10.59
400	7.39	8.05	11.70	10.3	17.3	11.4	15.3
600	11.2	12.20	18.20	16.2	25.3	17.2	22.6
800	15.3	16.6	24.8	22.1	34.3	22.1	30.2
1000	20.1	21.7	31.6	28.3	42.8	27.0	37.5
1200	25.1	26.9	38.7	37.6	51.6	32.2	45.1
1400	31.0	32.6	45.8	41.2	59.3	37.8	52.1
1600	37.0	38.5	52.1	47.6	65.8	43.9	58.5
1800	42.7	—	—	—	—	—	—
2000	48.3	—	—	—	—	—	—
2200	54.4	—	—	—	—	—	—
2400	60.2	—	—	—	—	—	—

Study [253] showed that, at temperatures $T < \theta_D$, the principal contribution to the anisotropy in resistivity comes from the anisotropy of a multi-connected surface of holes KM8h. For the value of $\rho_\perp / \rho_\parallel$ the following expression was obtained

$$\rho_\perp / \rho_\parallel = (a_\infty)^{-1}(1+\gamma/T^2)(1+\beta/T^2)^{-1}$$

where a_∞ is a temperature independent constant determined as

$$a_\infty = \int_{S_F^\perp} v_\perp \, dS_F^\perp \Big/ \int_{S_F^\parallel} v_\parallel \, dS_F^\parallel$$

where $v_\perp$ and $v_\parallel$ are the projections of the velocities of electrons in the directions perpendicular and parallel to the hexagonal axis, respectively; and, $dS_F^\parallel$ and $dS_F^\perp$ are the projections of corresponding elements of the Fermi surface. The integration was performed over the whole Fermi surface.

If the mean velocities $v_{av\perp}$ and $v_{av\parallel}$ are equal, the expression can be written as $a_\infty = S_F^\perp / S_F^\parallel$, where $S_F^\perp$ and $S_F^\parallel$ are the sums of areas from the projections of all of the Fermi surface sheets in the corresponding directions. The constants γ and β depend on the types of scattering processes in the region of closest approach between the neighboring sheets of the Fermi surface. This would be most effective for the Umklapp processes. In [253] it was found that for the temperature region of 95–160 K, the expression appears as $a_i = a_\infty + AT^{-2}$, where $a_\infty = 0.768$ and $A = 378$ K^2.

At temperatures $T \approx \theta_D$ the electron-phonon Umklapp processes are allowed

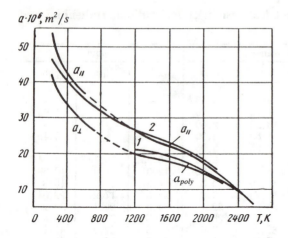

$a \cdot 10^6, m^2/s$

Figure 235 The diffusivity of ruthenium in the direction of the hexagonal axis ($a_\parallel$), perpendicular to it ($a_\perp$) and for polycrystals (a_{poly}) vs. temperature: *1*) [140, 260]; *2*) [39] for polycrystals.

for all parts of the Fermi surface; the wide angle scattering processes (scattering at 90°) are dominant, and the relaxation time is isotropic during phonon scattering. For this case $S_F^\perp/S_F^\parallel = a_\infty = \rho_\parallel/\rho_\perp$ and does not depend on temperature. The estimate of [257] produces $1.49 > \rho_\perp/\rho_\parallel > 1.09$, which describes the experimentally observed values of ρ for $\theta_D - 2\theta_D$. The decrease in the value of $\rho_\perp/\rho_\parallel$, at higher temperatures, is caused by the increase in the uncertainty in the Fermi parameters [140, 257]. Study [140] showed that the decrease in the anisotropy of resistivity depends linearly on the values of the average resistivity. Finally, it should be pointed out that the main contribution to resistivity and its anisotropy comes from the Umklapp processes (60–80% according to [257]) on parts of the Fermi surface that contain holes (mainly on the multiply-connected sheet).

Above 200 K, the diffusivity decreases with increases in temperature (Fig. 235) while, above 1200 K, the anisotropy in the diffusivity decreases [260]. The uncertainty of the data given in Fig. 235 and Table 44 is $\approx 5\%$.

Above 100 K, thermal conductivity and its electronic and phonon components

Table 44 The diffusivity values for ruthenium, rhodium, and palladium [39, 140] ($a \cdot 10^4$, m^2/s)

T, K	Ruthenium			Rhodium, poly	Palladium, poly	T, K	Ruthenium			Rhodium, poly	Palladium, poly
	poly	$\perp c$	$\parallel c$				poly	$\perp c$	$\parallel c$		
100	1.01	—	—	1.01	0.377	1200	0.260	0.198	0.256	0.283	0.255
200	0.450	0.42	0.55	0.562	0.261	1400	0.264	0.187	0.238	0.263	0.252
300	0.403	0.37	0.46	0.500	0.245	1600	0.233	0.174	0.220	0.246	0.242
400	0.389	0.33	0.40	0.465	0.246	1800	0.198	0.156	0.201	0.220	0.235
600	0.356	0.29	0.37	0.402	0.249	2000	0.184	0.140	0.174	0.190	—
800	0.319	0.25	0.33	0.345	0.251	2200	0.176	0.12	0.14	0.156	—
1000	0.290	0.22	0.29	0.308	0.254	2600	0.169	0.10	0.10	—	—

decrease with increases in temperature (Fig. 236). Below 500 K, the data were obtained for the same samples which were used in the investigation of electrical resistivity [259]. The data [260] also refer to the samples which were used for studying $\rho(T)$ above 300 K. For these data the uncertainty is about 8%. Overall thermal conductivity is determined by its electronic component. Below 1000 K, the value of λ_e^L is somewhat greater than λ_e^g, and, above 1000 K, the difference between these values is of the same magnitude as the uncertainty in the determination of λ and λ_g.

Above 200 K, the temperature dependence of the thermoelectric power of ruthenium has no inversion or extreme points. The value of S is negative and increases with temperature, which is typical for noble metals [55] (Fig. 237).

OSMIUM

At standard pressure, osmium has an hcp structure which is stable up to the very high melting point $T_m = 3283 \pm 10$ K [39]. At room temperature, osmium has the lattice parameters $a = 0.2733$ nm, $b = 0.43195$ nm [82].

Osmium, as ruthenium, has a relatively small density of electronic states (see

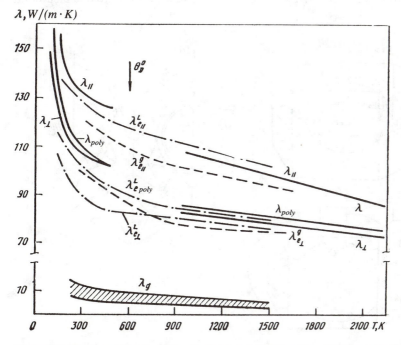

Figure 236 The thermal conductivity of ruthenium in the direction of the hexagonal axis ($\lambda_{\parallel}$), perpendicular to it ($\lambda_{\perp}$) and for a polycrystalline sample (λ_{poly}); below 500 K the data are from [259], above 100 K, from [260].

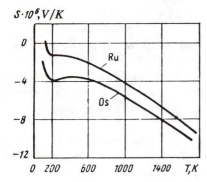

$S \cdot 10^6, V/K$

Figure 237 The thermoelectric power (S) of ruthenium and osmium vs. temperature [55].

Fig. 93). Studies [41, 256] showed that the structure of the Fermi surface of osmium has much in common with that of ruthenium. The electronic ellipsoids at the Γ point, in particular, are similar to those of ruthenium (Fig. 238), although they are larger and, at the A point, come closer to the boundary of the Brillouin zone.

Table 42 gives the data for polycrystalline osmium; the elastic moduli are record high. The electrical resistivity of single crystals of osmium is given in Fig. 239 and Table 43. Above room temperature, the data are taken from [257] for crystals with $r \approx 20$. Below 293 K, the data are new. They were obtained in the

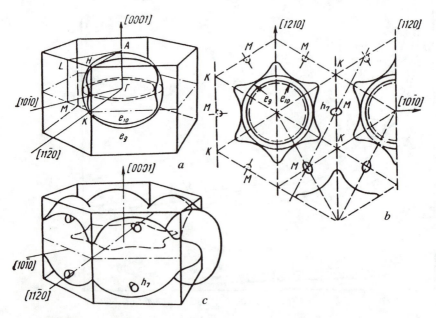

Figure 238 The Fermi surface of osmium [256]: *a*) the electronic sheets e_9 and e_{10}; *b*) the cross section through the Γ point in the direction perpendicular to (0001); *c*) the surface of holes h_8 and the small ellipsoids h_7.

laboratory of N. V. Wolkenshtein for samples with $r \approx 2000$. The uncertainty in the given values is $\approx 2\%$. The anisotropy in the electrical resistivity reaches 1.5 at 1300–1500 K, and then it decreases. The anisotropy in the diffusivity behaves similarly. Study [257] suggested that the main contribution to resistivity, ρ_h, is produced by the carriers of the surface of holes, which largely determines the anisotropy of the electrical resistivity. At Debye temperature, $\rho_\perp/\rho_\parallel = S_\parallel/S_\perp$ and for the surface of holes $S_\parallel/S_\perp = 1.60$ [140, 256]. The calculations performed in [257] show that, at high temperatures, the principal contribution to the scattering processes is coming from the Umklapp processes (up to 70%). The decrease in the anisotropy above 1000 K was explained in [257] by the increased uncertainty in the Fermi impulse. This is because the mean free path of carriers decreases and approaches the interatomic distance. This factor also explains the convergence of the corresponding kinetic coefficients of ruthenium and osmium at high temperatures. The increase in the uncertainty of the Fermi impulse makes the difference between the Fermi surfaces of these metals unnoticeable.

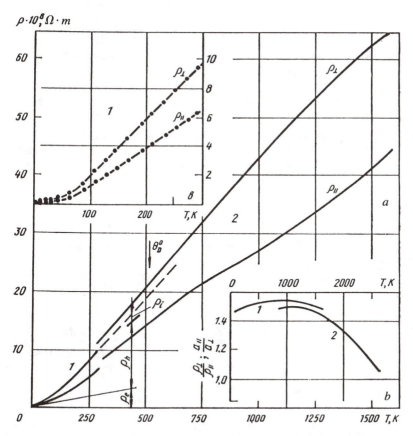

Figure 239 The electrical resistivity of osmium in the direction of the hexagonal axis ($\rho_\parallel$) and perpendicular to it ($\rho_\perp$) vs. temperature: _1_) the data of N. V. Wolkenshtein for a sample with $r \approx 2000$; _2_) [257]; in the inset the anisotropy of _1_) resistivity; _2_) diffusivity vs. temperature.

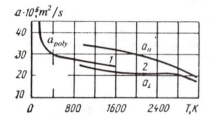

Figure 240 The diffusivity of osmium in the direction of the hexagonal axis ($a_\parallel$) and perpendicular to it ($a_\perp$) vs. temperature: *1*) [39], for a single crystal; *2*) [140].

The temperature dependence of the diffusivity is given in Fig. 240 and Table 45. The data [257] were obtained with the same single crystal which was used for investigating the electrical resistivity. The uncertainty in the given values is ≈5%. As temperature approaches the melting point, the anisotropy in diffusivity decreases (see Fig. 240). This phenomenon has the same explanation as in the electrical resistivity. The temperature dependence of the thermal conductivity of osmium is unusual because the signs of the derivatives depend on the crystallographic direction: $\partial\lambda_\parallel/\partial T < 0$, and $\partial\lambda_\perp/\partial T > 0$ [257]. This is the result of peculiarities of the electron scattering in osmium. Above 800 K, different signs were observed for the second derivatives of the electrical resistivity with respect to temperature: $\partial^2\rho_\parallel/\partial T^2 > 0$, and $\partial^2\rho_\perp/\partial T^2 < 0$. The thermal conductivity of osmium is largely electronic in nature. Figure 241 shows that, below 1500 K, some difference between λ_e^L and λ_e^g exists, hence $L > L_0$, but, above 1500 K, this difference does not exceed the uncertainty in the determination of λ (10%).

The temperature dependence of the thermoelectric power of osmium is similar to that of ruthenium, and, at high temperatures, the absolute values of S are observed to converge (see Fig. 237).

Table 45 The diffusivity values for osmium, iridium, and platinum ($a \cdot 10^4$, m²/s) [39, 140]

T, K	Osmium			Iridium	Platinum	T, K	Osmium			Iridium	Platinum
	poly	⊥ c	‖ c				poly	⊥ c	‖ c		
40	9.38	—	—	14.8	1.65	1400	0.253	0.223	0.327	0.322	0.266
80	1.54	—	—	1.27	0.437	1600	0.245	0.218	0.318	0.300	0.257
100	0.93	—	—	0.842	0.359	1800	—	0.214	0.304	0.280	0.244
200	0.353	—	—	0.558	0.268	2000	—	0.214	0.299	0.266	0.222
300	0.300	—	—	0.502	0.252	2200	—	0.214	0.285	0.251	—
400	0.294	—	—	0.482	0.247	2400	—	0.214	0.268	0.232	—
600	0.286	—	—	0.451	0.244	2600	—	0.209	0.258	0.20	—
800	0.277	—	—	0.420	0.244	2800	—	0.201	0.237	—	—
1000	0.269	0.237	0.346	0.380	0.248	3000	—	0.187	0.215	—	—
1200	0.261	0.230	0.337	0.349	0.260						

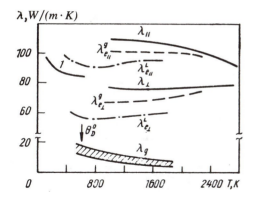

Figure 241 The thermal conductivity of osmium in the direction of the hexagonal axis ($\lambda_\parallel$) and perpendicular to it ($\lambda_\perp$) vs. temperature: *1*) λ_{poly} [259].

§9 RHODIUM, IRIDIUM

RHODIUM

At standard pressure, rhodium has an fcc structure with the lattice parameter $a = 0.38044$ nm at 293 K. The Fermi surface of rhodium has been investigated in studies [41, 261]. It consists of two big electronic surfaces at the Γ point, two pockets of holes at the X point and a small closed surface of holes at the L point.

The curve of the density of electronic states $N(\varepsilon)$ has been investigated in detail [69, 70, 261, 262]. It was found that the Fermi level of rhodium lies near the minimum of $N(\varepsilon)$ (see Fig. 87). This makes rhodium a metal typical of the so-called 'plus' group, the kinetic properties of which exhibit a specific behavior at high temperatures.

Table 42 gives the elastic characteristic of polycrystalline rhodium at $T = 293$ K. The temperature dependence of the electrical resistivity of rhodium $\rho(T)$ is presented in Fig. 242 and also below [140].

T, K	100	200	300	400	600	800
$\rho \cdot 10^8$, $\Omega \cdot m$	0.91	2.95	5.01	7.10	11.50	16.0

T, K	1000	1200	1400	1600	1800	2000
$\rho \cdot 10^8$, $\Omega \cdot m$	20.8	26.0	31.1	36.9	42.0	48.0

The information on the kinetic properties of rhodium applies to a metal of $\approx 99.9\%$ purity. The uncertainty in the given values is 5%; above 1500 K the data require verification. The dependence of $\rho(T)$ is characterized by a positive deviation from linearity ($\partial^2 \rho / \partial T^2 > 0$).

Study [11] linked the kinetic properties of rhodium with its electronic struc-

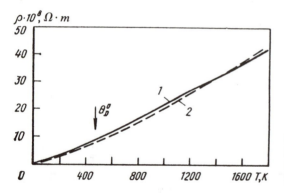

Figure 242 The electrical resistivity of rhodium (ρ) as a function of temperature: *1*) [14, 217, 259]; *2*) calculations using Mott's *s-d* model [11].

ture. It calculated the temperature dependences of the electrical resistivity, thermal conductivity and thermoelectric power of rhodium using the two-band model of electron-phonon scattering and compared the results with the existing data. Close agreement between the experimental dependence of $\rho(T)$ and the theoretical estimates (see Fig. 242) allowed the authors of [11] to conclude that, within the framework of the *s-d* model, one-to-one relationships exist between the dependences of $\rho(T)$ and $\varkappa(T)$ and the density of the electronic states.

Table 44 shows the diffusivity of rhodium [39, 140]. The values decrease over the entire temperature interval, and for temperatures above 1600 K the rate of this reduction decreases. The uncertainty in the given values is $\approx 5\%$.

Figure 243 shows the temperature dependence of the thermal conductivity of rhodium; the uncertainty of the data is 10%. Figure 243 also shows the results of estimating the electronic component on the basis of the W-F-L law, from $\lambda_e^L = L_0 T/\rho$, and from calculating the lattice component $\lambda_e^g = \lambda - \lambda_g$. Above 1000 K, these results coincide within the uncertainty of the calculation. Simliar results were also obtained by calculating the temperature dependence of $\lambda_{e\text{-}d}^L$ [11] from the *s-d* model.

Thus, for rhodium, the dependence of $\rho(T)$, when $\partial^2\rho/\partial T^2 > 0$, correlates with the dependence of $\lambda(T)$, when $\partial\lambda/\partial T < 0$. This is because the mechanisms of scattering and the relaxation time are the same for both properties, and the

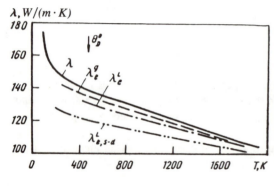

Figure 243 The thermal conductivity of rhodium (λ) as a function of temperature [140, 217, 259], $\lambda_e^g = \lambda - \lambda_g$; λ_g is the calculation from [23]; $\lambda_{e\text{-}d}^L$ is the calculation based on Mott's *s-d* model [11]; $\lambda_e^L = L_0 T/\rho$.

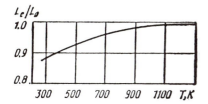

Figure 244 The reduced electronic Lorenz function of rhodium (L) [140].

lattice contribution to the thermal conductivity is no more than a small percentage of the total value. At temperatures below 600 K the processes of inelastic scattering of electrons become appreciable; this is evident from the analysis of the temperature dependence of the Lorenz function (Fig. 244). Such a dependence of L/L_0 is typical for metals with a small density of electronic states [11, 23, 140].

Figure 245 shows the temperature dependence of the absolute thermoelectric power of rhodium [55, 263]; the values are small and negative. It is interesting that calculation [11] produces a very different temperature dependence, demonstrating the difficulties in the application of the 'band' approach to the analysis of this property in transition metals.

IRIDIUM

At standard pressure, iridium has an fcc structure with the lattice parameter $a = 0.38388$ nm [74].

The structure of the Fermi surface of iridium is similar to that of rhodium [41, 261] although the small pockets near the L point are absent. Similar to rhodium, the Fermi level of iridium lies near the minimum on the curve of the density of electronic states [41, 70, 261, 262].

The information on the elastic characteristics of polycrystalline iridium at 293 K is given in Table 42. At 300 K, single crystals of iridium have the following elastic constants ($c_{ij} \cdot 10^{-11}$, Pa): $c_{11} = 5.800$, $c_{12} = 2.420$, $c_{44} = 2.560$ [264].

The temperature dependence of the electrical resistivity of iridium is similar to that of rhodium (Fig. 246) [82, 259]. The uncertainty in the data given in Fig. 246 and Table 43 is about 4%; the data apply to a metal of 99.8% purity. The calculation of the temperature dependence of resistivity was made using Mott's

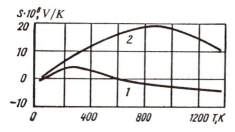

Figure 245 The absolute thermoelectric power of rhodium (S) vs. temperature: *1*) the experimental data of [55, 263]; *2*) the calculations using Mott's *s-d* model [11].

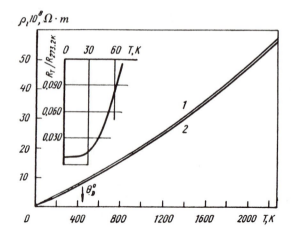

Figure 246 The electrical resistivity of iridium (ρ) as a function of temperature: *1*) [82, 217, 259]; *2*) the calculation based on Mott's *s-d* model [11]. In the inset: the data obtained in the laboratory of professor N. V. Volkenshtein.

s-d model and taking into account the parameters for four sheets of the Fermi surface. It produced a satisfactory explanation for the observed slope of ρ(T), when $\partial^2 \rho / \partial T^2 > 0$ [11].

The diffusivity of iridium is given in Table 45 [140, 265]. The data apply to a metal of 99.9% purity with $r = 90$. The uncertainty in these values is about 4% below 2400 K and about 6% at higher temperatures. The dependence of a(T) for iridium, as for rhodium, is characterized by a negative temperature coefficient above 100 K.

The experimental data on the thermal conductivity of iridium of 99.9% purity, taken from studies [48, 140, 217, 259], are shown in Fig. 247. The uncertainty in their determination is ~8%. At high temperatures, the temperature coefficient of the thermal conductivity of iridium is negative, as was the case for rhodium. It correlates with the positive curvature in the dependence of λ(T). Table 46 lists the data on the thermal conductivity of iridium.

The values of the thermoelectric power of iridium are characterized by relatively small magnitudes; above 400 K they become negative and their magnitude increases with temperature (Fig. 248).

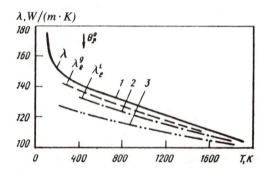

Figure 247 The overall (λ) and electronic (λₑ) thermal conductivity of iridium vs. temperature: *1*) the experimental data of [48, 140, 217, 259]; *2*) the electronic thermal conductivity $\lambda_e^L = L_0 T / \rho$; *3*) the theoretical values of the electronic thermal conductivity calculated by using Mott's model [11].

Table 46 The thermal conductivity of ruthenium, rhodium, iridium and osmium (λ, W/(m · K)) [82, 140, 217, 259]

T, K	Ruthenium			Osmium			Rhodium	Iridium
	$\perp c$	$\parallel c$	poly	$\perp c$	$\parallel c$	poly		
100	140	180	150	—	—	—	185	—
150	123	150	132	—	—	96	160	—
200	116	138	125	—	—	93	156	151
250	112	135	121	—	—	91	154	149
273	110	134	119	—	—	88	153	148
300	108	133	117	—	—	87	152	147
400	105	129	115	—	—	86	145	141
600	95	120	105	—	—	85	135	136
800	87	112	96	—	—	—	126	130
1000	81	106	89	75.5	110	—	121	125
1200	77	101	83	75.0	110	—	118	121
1400	74	97	80	74.5	109	—	115	117
1600	73	93	77	74.5	108	—	109	110
1800	71	89	74	75	106	—	106	106
2000	67	83	70	76	105	—	104	103
2200	—	—	—	78	104	—	—	—
2400	—	—	—	79	99	—	—	—
2600	—	—	—	78	96	—	—	—
2800	—	—	—	76.5	90	—	—	—

Aisaka and Shimizu [11] calculated the temperature dependences of the electrical resistivity, electronic thermal conductivity and thermoelectric power of iridium using Mott's model (Fig. 246). As was the case for rhodium, the Mott's model turned out to be successful in describing the electrical and thermal properties of iridium. This proves that, at high temperatures, the electron-phonon interband scattering processes are dominant. However, the theoretical and experimental curves of $S(T)$ differ substantially for iridium, as well as for rhodium. This indicates that Mott's model is not applicable to properties which are sensitive to the characteristics of the electronic spectrum.

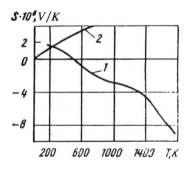

Figure 248 The thermoelectric power of iridium (S) as a function of temperature: *1*) the experimental data of [55]; *2*) the calculation from Mott's model [11].

§10. PLATINUM, PALLADIUM

PALLADIUM

At standard pressure, palladium has an fcc structure with the lattice parameter a = 0.3883 nm at 293 K. Palladium is a paramagnetic metal belonging to the $4d$-group. The $4d$- and $5f$-bands overlap considerably in this metal, therefore about 0.36 electrons per atom transfer from the d-band into the s-band [266]. As a result, the d-band becomes partially vacant, giving rise to the properties of palladium, that are typical for the transition metals.

The electronic structure of palladium is known quite well [41, 261, 266, 267]. The data on the electronic heat capacity and magnetic susceptibility of palladium [268, 269] show that the Fermi level of palladium lies near the maximum of the density of states; this governs the specific physical properties of palladium (Fig. 249).

The experimental and theoretical investigation of magnetic susceptibility, and the information on the Knight shift [270] permitted the conclusion that palladium belongs to the transition metals with a strong exchange interaction.

The elastic constants of palladium are shown in Fig. 250; below 300 K the data are taken from [271], above 300 K from [272]. As in metals of the Fifth group, the anomaly in c_{44} near 150 K is related to the specific features of the band structure. Figure 251 shows the elastic moduli of single crystals of palladium of 99.91% purity [273].

The following relation, which is typical for cubic crystals, holds for palladium: $E_{\langle 111 \rangle} > E_{\langle 110 \rangle} > E_{\langle 100 \rangle}$. At room temperature, the ratio $E_{\langle 111 \rangle}/E_{\langle 110 \rangle} = 2.64$ is similar to that of other fcc metals, and it increases with temperature. Figure 251 also shows Young's modulus of polycrystalline palladium. The value of E_{poly} at $T = 293$ K differs by 1.5% from the result of [73], given in Table 42.

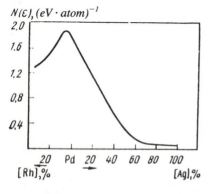

Figure 249 The density of the electronic states of rhodium, palladium, and silver as a function of energy [261, 262].

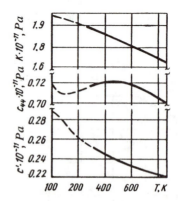

Figure 250 The elastic constants of palladium as a function of temperature [271, 272].

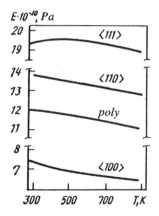

Figure 251 Young's modulus (E) of palladium as a function of temperature [273].

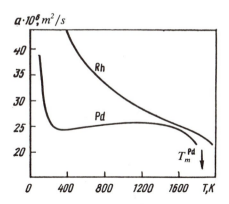

Figure 252 The diffusivity of palladium and rhodium (a) vs. temperature [39, 140].

The information on the electrical resistivity of palladium is compiled in study [63]. The main distinguishing feature in the dependence of $\rho(T)$ is its negative curvature above 100 K.

The resistivity of palladium at different temperatures is listed below [140].

T, K	50	100	200	300	400	600
$\rho \cdot 10^8$, $\Omega \cdot$ m	0,606	2,634	6,887	10,804	14,46	21,09

T, K	800	1000	1200	1400	1600
$\rho \cdot 10^8$, $\Omega \cdot$ m	26,89	31,92	36,4	40,4	45

The temperature dependence of the diffusivity of palladium is different from that of rhodium; near 1000–1400 K a wide maximum is observed. The data given in Fig. 252 and Table 44 refer to palladium of 99.99% purity; the uncertainty of the data is $\approx 5\%$.

The thermal conductivity of palladium is taken from compilation [140]. The most likely appearance of the temperature dependence of $\lambda(T)$ is shown in Fig. 253, and the values are given below [140]. This data refer to a metal of 99.99% purity; the uncertainty is 6%.

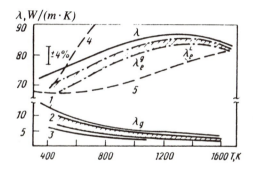

Figure 253 The thermal conductivity of palladium (λ) vs. temperature: 1–3) the lattice component, calculated from the data of various authors [140]; 4) the calculated values of λ_e [11]; 5) $\lambda_e^L = L_0 T/\rho$.

T, K	100	200	300	400	600	800	1000	1200	1400	1600
λ, W/(m·K). . .	76.0	75.0	75.2	75.5	79.0	83.0	87.0	86.2	86.9	86.0

The temperature dependence of the thermal conductivity of palladium has a wide minimum near the room temperature and a weak maximum at 1200–1300 K. The preliminary analysis of this dependence shows that it is of an electronic nature, although the difference between λ_e^L and λ_e^g becomes noticeable for temperatures below 1400 K.

The temperature dependence of the absolute thermoelectric power of palladium is given in Fig. 254. For temperatures above 200 K the values of S and $\partial S / \partial T$ are high and negative.

The kinetic properties of palladium have much in common with those of platinum; therefore, the special features of the scattering mechanisms are analyzed together for both metals.

PLATINUM

Platinum, as palladium, has an fcc structure with the lattice parameter $a = 0.39239$ nm at 298 K. The Fermi level of platinum lies to the right of the maximum density of electronic states (see Fig. 93); but the values of $N(\varepsilon)$ are somewhat smaller than those of palladium. Similar to palladium, platinum has an enforced exchange interaction [69, 275, 276]. The structure of the Fermi surface of platinum is similar to that of palladium [41].

Figure 255 gives the elastic constants of platinum. Similar to palladium, vanadium, niobium and tantalum, the anomalous behaviour of the elastic constant c_{44} is caused by the specific features of the electronic spectrum of platinum [277].

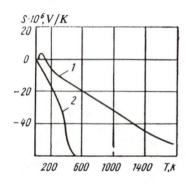

Figure 254 The absolute thermoelectric power of palladium (S) vs. temperature: *1*) [274, 55]; *2*) the calculations of [11].

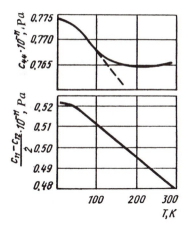

Figure 255 The elastic constants of platinum as a function of temperature [277].

From 300 to 1250 K Young's and shear moduli of platinum are linear functions of temperature. They can be described by equations [238]:

$$E = (1.68 - 3.38 \cdot 10^{-4}T \pm 0.046) \cdot 10^{11} \text{ Pa}$$

$$G = (0.624 - 1.15 \cdot 10^{-4}T \pm 0.015) \cdot 10^{11} \text{ Pa}$$

The measurements were made on a sample annealed at 1000° C for 20 hours. The values of moduli at 293 K differ from those given in Table 42 by 2–3%.

Platinum is a standard thermometric material. There are detailed tables for the temperature dependence of its electrical resistivity below 300 K [100]. The data for temperatures above 300 K are given below. They correspond to the average data of [259, 281–286]; the uncertainty in the determination of the data is ≈2%:

T, K	50	100	200	300	400	600
$\rho \cdot 10^{8}$, $\Omega \cdot$m	0.736	2.795	6.884	10.81	14.60	21.85

T, K	800	1000	1200	1400	1600	1800
$\rho \cdot 10^{8}$, $\Omega \cdot$m	28.70	35.10	40.89	46.69	51.80	56.52

The kinetic properties of platinum are usually studied using samples of 99.99% purity. For platinum, the temperature dependence of resistivity, as in the case of palladium, is characterized by a negative curvature above 100 K (Fig. 256).

The information on the diffusivity of platinum is generalized in studies [39, 140]. Table 45 lists the average values of $a(T)$. It is interesting to note that, for platinum, the shape of the dependence of $a(T)$ and, even the values, are close to those of palladium. The comparison of the dependence with that of iridium reveals the same features as were found for palladium: $a(T)$ of platinum has a weak max-

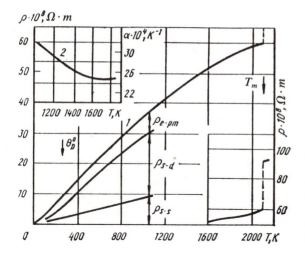

Figure 256 The electrical resistivity of platinum (ρ) vs. temperature: *1*) below 300 K [100]; above 300 K, the average data of [263, 281, 283–285]; *2*) the temperature coefficient of the electrical resistivity [286]; for the liquid state [227].

imum at high temperatures, as compared to the monotonically decreasing values of $a(T)$ for iridium.

The thermal conductivity of platinum at moderate and high temperatures was compiled in studies [39, 48, 140, 281–284]. The most probable values of λ are given below; their uncertainty is $\approx 3\%$ below 700 K and about 5% at higher temperatures (Fig. 257). The thermal conductivity of platinum, as in palladium, increases with temperature.

T, K	100	200	300	400	600	800
λ, W/(m·K)	85,0	79,0	74,1	73,2	73,0	74,8

T, K	1000	1200	1400	1600	1800	2000
λ, W/(m·K)	76,9	83,2	88,3	89,5	89,0	86,0

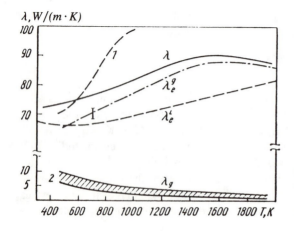

Figure 257 The thermal conductivity of platinum (λ) vs. temperature [39, 140]: *1*) the calculation of λ_e from [11]; *2*) λ_g according to the estimate of [23].

Platinum is a thermometric material, therefore its thermoelectric power is often used to determine that of other metals. The absolute thermoelectric power of platinum is given in Fig. 258 and below [82, 100].

T, K	18	33	73	113	153	193	233	273
$S \cdot 10^6$, V/K	+1.81	+4.04	+5.88	+3.56	+1.18	−0.92	−2.78	−4.42

T, K	293	373	473	573	673	773	873
$S \cdot 10^6$, V/K	−5.13	−7.3	−9.2	−10.9	−12.5	−14.0	−15.6

T, K	973	1073	1173	1273	1473	1673	1873
$S \cdot 10^6$, V/K	−17.1	−19.6	−20.1	−21.6	−24.6	−27.6	−30.6

The Fermi levels of palladium and platinum are located near the maximum of the density of states; therefore, the calculations were initially based on Mott's model [2]. Later, this model was frequently used in a number of studies, the best of which were the ones by Aisaka and Shimizu [11] and by Laubitz and Matsumura [279]. These have been referenced here many times. The results of both studies agree fairly well, even though the former authors took into account not only the dependence of $N(\varepsilon)$, but also the effective masses of the carriers in the s- and d-bands, the parameter of the electron-phonon interaction, and the temperature dependence of the Debye temperature caused by thermal expansion. The studies describe quite well the dependence of the electrical resistivity on temperature. However, in the more accurate study [11], the values of $\rho(T)$ calculated for palladium differ from the experimental data more than do the values of [279].

Figure 253 shows the electronic component of the thermal conductivity, and the Lorenz number of palladium. The values of λ_e isolated by using the calculated lattice component of conductivity [23, 140, 279, 280] are also shown in Fig. 253. It is evident that all of the calculated values of λ_g practically coincide, giving a common band of possible values amounting to 5–12% of the value of λ at 500 K and only 3–5% at 1000 K. For platinum, analogous estimations of λ_g and λ_e

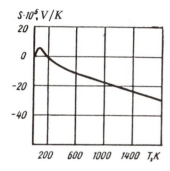

Figure 258 The absolute thermoelectric power of platinum (S) vs. temperature [263].

are given in Fig. 257. It should be pointed out that the values of λ_e obtained by using the estimates of λ_g practically coincide, at 300 K and near the melting point, with the values calculated by using the standard W-F-L law. However, at intermediate temperatures, the difference between them substantially exceeds the possible uncertainty limit and reaches a maximum near 1100 K for palladium, and approximately 1500 K for platinum.

Figure 259 shows the temperature dependences of the reduced Lorenz functions for palladium and platinum, obtained in different ways. The number *1* curves correspond to the dependence $L/L_0 = \lambda - \lambda_g/\lambda_e^L$, where $\lambda_e^L = L_0 T/\rho$. The number *2* curves correspond to the more accurate values of the Lorenz function that have been corrected for the thermoelectric power $L/L_0 = (\lambda - \lambda_g)/\lambda_e^L - S^2/L_0$. The difference between these functions is small, even for palladium which has the highest value of the thermoelectric power among the considered metals. This proves that the difference between the results of the two calculations can be neglected for most metals.

Figure 259 also shows the values of L/L_0 vs. temperature, obtained in [11] and [279]. These results show a qualitatively correct relation: above 500 K $L > L_0$, however the magnitude of these values differs considerably from the most reliable results described by the curves *1* and *2*.

Finally, Fig. 259 shows the calculated values of L/L_0 versus temperature [280]. The accuracy of these calculations is probably higher than of previous ones. The results show the increase in the Lorenz number with temperature, but they do not show the observed tendency of $L(T)$ to assymptotically approach the standard value L_0 near the melting point.

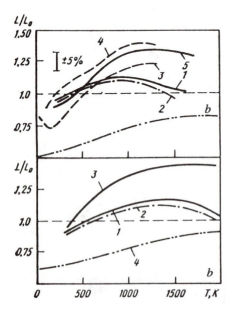

Figure 259 The reduced Lorenz functions (L/L_0) for palladium and platinum [140] vs. temperature: a) for palladium: *1*) $L/L_0 = (\lambda - \lambda_g)/\lambda_e^L$; *2*) the same as in *1*), but with a correction for the thermoelectric power; *3*) L/L_0 [280]; *4*) L/L_0 [279]; *5*) L/L_0 [11]; *6*) L/L_0 [17] with $S_{St} = 10$ and $T_{F_i} = 2000$ K; b) for platinum: *1*) $L/L_0 = (\lambda - \lambda_g)/\lambda_e^L$; *2*) $L/L_0 = (\lambda - \lambda_g)/\lambda_e^L - S^2/L_0$; *3*) L/L_0 [11]; *4*) L/L_0 [17] with $S_{St} = 5$ and $T_{F_i} = 2000$ K.

Recently, Fradin, et al. calculated the electrical resistivity of palladium [282] and platinum [275] at high temperatures using a computer program containing fairly detailed information on $N(\varepsilon)$ and the Fermi surface of these metals. These calculations showed that the inter-band scattering provides a satisfactory explanation for the specific high-temperature behaviour of $\rho(T)$ and $\varkappa(T)$. Unfortunately, the authors did not make similar calculations for the Lorenz functions. However, it can be assumed that their results would not be much different from the data of [280] and [11] because Fradin, et al. used the same model.

On the other hand, there recently appeared a number of studies which calculated the values of $\rho(T)$, based on the paramagnon model [16, 17]. In [16] the dependence of $\rho(T)$ for palladium was adequately described using three parameters: the width of the d-band ($T_{F_i} = 2000$ K), the Stoner parameter ($S_{St} = 10$) and the ratio of the mean quasi-impulse of s-electrons to that of all electrons. However, contrary to the data of [16], Fradin [282] showed that the dependence of $\rho(T)$ for platinum can be explained by selecting, apriori, the d-band type to be within the inter-band scattering framework without having to resort to spin fluctuation scattering. Furthermore, using the calculation of $N_d(\varepsilon)$ for platinum, study [282] gave a good description of $\rho(T)$.

The question of which of the two models is the best can be resolved by comparing other kinetic properties calculated from these models, particularly thermal conductivity and the Lorenz number.

Figure 259 shows the results of calculating the dependence of $L/L_0(T)$ [140] from the same parameters Jullien, et al. [16] used for palladium in conjunction with the Beal-Monod and Mills formula [17], based on the same paramagnon model (curve 6 for palladium and 4 for platinum). For both metals, the deviation of L-L_0 is opposite to that observed experimentally. They tend, however, to approach each other near the melting points.

The final conclusion about the possible mechanisms of high-temperature electron scattering is that $\rho(T)$ can be described quite successfully by either the two-band model, accounting for the Mott s-d transitions, or by the paramagnon model. The first model, in calculating the Lorenz number and thermal conductivity overestimates the values in comparison with L_0; the overestimation increases with temperature. The second model underestimates the values of L/L_0 making them approach to one. The comparison of the calculations with the experimental data showed that neither of the two models can adequately describe the dependences of $L(T)$ and $\lambda(T)$. It seems, however, that their combination may give a good result because the differences between the values of ΔL predicted by the two models and the experimental values have opposite signs.

Considering the previous results, one more important conclusion can be drawn from Figure 259. The band calculations lead to a significant overestimation of L/L_0 for $T \rightarrow T_m$. At the same time $L/L_0 \rightarrow 1$, indicating that the model becomes less applicable as T approaches T_m. The limitations due to the decrease in the mean free path of carriers should be mentioned.

The Hall coefficients for the metals of the platinum group are listed below [92, 100].

Ruthenium:

T, K 293
$R \cdot 10^{10}$, m³/C

Ru_{poly} . . . . +2.2
$\perp c$ . . . . +1.77

$\parallel c$ . . . . +1.14

Platinum:

T, K	83	173	273	573	873
$R \cdot 10^{10}$, m³/C	−0.203	−0.181	−0.214	−0.253	−0.278

Rhodium:

T, K 291
$R \cdot 10^{10}$, m³/C . . . . +4.9

Palladium:

T, K	299	337
$R \cdot 10^{10}$, m³/C	−0.845	−0.844

§11 IRON, COBALT, NICKEL

IRON

At standard pressure and temperatures below 1183 K, iron has a bcc crystalline structure with the lattice parameter $a = 0.28664$ nm at 293 K. Below the Curie point (1042 ± 0.5 K) [39, 82], this modification is usually called α-Fe; and, in the paramagnetic region up to 1183 K, β-Fe. γ-Fe has an fcc structure with the lattice parameter $a = 0.36468$ nm at 1189 K. The γ-δ transformation occurs at 1667 K; δ-Fe again has a bcc structure with the lattice parameter $a = 0.29322$ nm.

The electronic structure of iron has been studied mainly for the ferromagnetic state [41]. The experimental studies of the Fermi surface of iron are insufficient for the ferromagnetic state, and, for the paramagnetic high-temperature state no such investigations have been performed at all. The parameters of the band structure for paramagnetic and ferromagnetic iron were theoretically studied in [287].

In the temperature interval from 300 to 1170 K, the temperature dependence of the elastic constants of iron is essentially nonlinear. In the ferromagnetic phase below and above $T_c = 1043$ K, the dependence of $c_{ij}(T)$ is influenced by the structural transformation ($T_{\beta\text{-}\gamma} = 1183$ K). A small λ-type anomaly in the longitudinal elastic constant c_{11} at T_c is typical for the magnetic phase transitions of the second order. Among the shear elastic constants, the highest changes near the magnetic ordering are experienced by $c' = (c_{11} - c_{12})/2$. The difference between c' and c_{44} in sensitivity towards the magnetic changes justifies the assumption that the exchange interactions exist mainly between the second, rather than the first, closest neighbors.

The elastic constants of iron ($c_{ij} \cdot 10^{-11}$, Pa) are given below [288].

T, K	298	373	473	573	673	773	873	923	973
c_{11}	2.322	2.277	2.215	2.142	2.056	1.968	1.867	1.808	1.738
c'	0.483	0.467	0.441	0.411	0.380	0.344	0.301	0.278	0.248
c_{44}	1.170	1.154	1.132	1.111	1.091	1.072	1.053	1.042	1.031

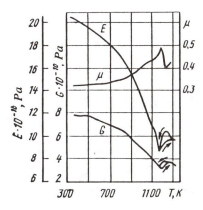

Figure 260 Young's modulus (E), shear modulus (G), and Poisson's coefficient (μ) of iron vs. temperature [289].

Continuation

T, K	1013	1033	1043	1053	1073	1133	1173
c_{11}	1.667	1.626	1.598	1.597	1.581	1.523	1.488
c'	0.219	0.199	0.187	0.181	0.170	0.146	0.133
c_{44}	1.021	1.016	1.013	1.010	1.005	0.996	0.990

Figure 260 shows the elastic moduli of polycrystalline iron for the temperature interval from 300 to 1300 K. A sharp increase in the curvature of $E(T)$ is observed for the region of the ferromagnetic ordering. The β-γ structural transformation is marked in the temperature dependence of the shear and Young's moduli by minima, and in $\mu(T)$ by a maximum. The hysteresis during this transformation is natural [289].

The electrical resistivity of iron and its derivative with respect to temperature have been investigated by many authors [15, 140, 290–294] (Figs. 261, 262); the uncertainty in the given values is 1–2%. The anomalies near the phase transformation points are well pronounced; there are jumps in the values of $\rho(T)$ during structural transformations and an inflection point during the magnetic disordering transformation. A certain hysteresis is observed for the value of $\rho(T)$ near the structural transformations; it is determined by the rate of cooling and heating. Impurities make the localization of the transformations less pronounced. Figure 261 shows the temperature dependence of the electrical resistivity near the Curie point [294]. The experimentally obtained data are described by the equations

$$\partial\rho/\partial T = 0.84 - 0.76 t^{\alpha} \quad \text{at} \quad T < T_C$$

$$\partial\rho/\partial T = 0.72 - 0.76 t^{\alpha} \quad \text{at} \quad T > T_C$$

where $t = (T - T_c)/T_c$ is the reduced temperature; $\alpha = 0 \pm 0.1$; $\partial\rho/\partial T$ is $\mu\Omega \cdot \text{cm/K}$.

The theoretical calculations, based on the theory of similarity, show that the anomaly in $\partial\rho/\partial T$ should be similar to the anomaly in the specific heat with the same critical index α [15, 27–29, 294].

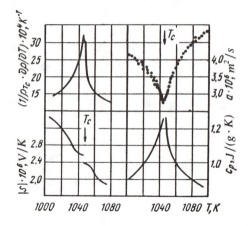

Figure 261 The temperature dependences of the temperature coefficient of electrical resistivity $(1/\rho_{T_c} \cdot \partial\rho/\partial T)$ [294], diffusivity (a) [295], thermoelectric power (S) [296] and the specific heat (C_p) [295] for iron near the Curie point.

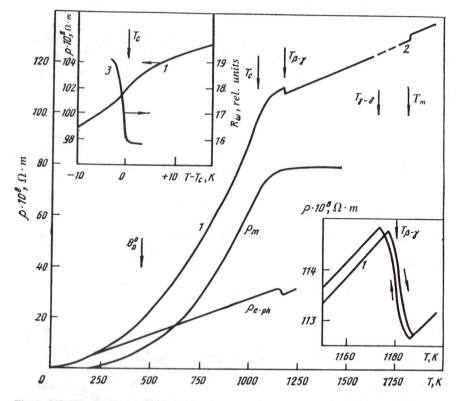

Figure 262 The electrical resistivity of iron (ρ) vs. temperature: *1*) the average data of [140, 290–292, 294] (below 300 K the data of [15] are taken into account); *2*) [297]; *3*) [294] for the frequency of 1500 Hz near the Curie point (R_ω, relative units).

Study [293] discussed the influence of impurities on the electrical resistivity and thermal conductivity of iron, and also determined the correlation between the electrical resistivity and thermal conductivity values.

The diffusivity of iron (Fig. 263) has a large negative temperature coefficient below the Curie point, a minimum in the region of T_c, and a jump-like change at the β-γ and γ-δ transformations. The anomaly in the diffusivity near T_c is described by the equations

$$a = A |t|^{\gamma - \alpha} \approx A |t|^{\gamma}$$

where $\gamma = 0.16 \pm 0.04$ at $T < T_c$ and $\gamma = 0.17 \pm 0.04$ at $T > T_c$.

The uncertainty in the data given in Fig. 261 and at the end of this section is about 4–5% below 1600 K and about 10% at higher temperatures.

The temperature dependence of the electrical resistivity of iron has a typical appearance for a ferromagnetic material: $\rho = \rho_{e-ph} + \rho_{e-e}$, where ρ_{e-ph} is the electron-phonon, and ρ_{e-e} is the electron-electron components. The latter is mainly

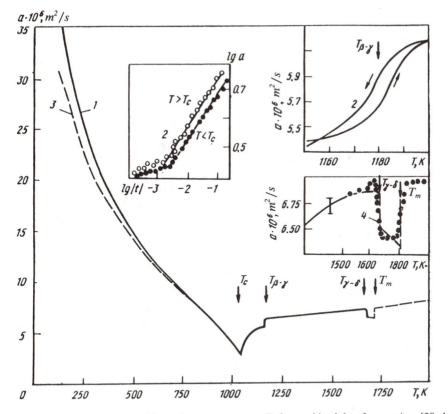

Figure 263 The diffusivity of iron (a) vs. temperature: 1) the combined data for pure iron [39, 140]; 2) [140, 295] for the anomalies in a(T) near the phase transformation points; 3) [39] for Armco iron; 4) the new data of the present author.

of a magnetic nature, $\rho_{e-e} = \rho_m$ [15, 140]. In the ferromagnetic region, the temperature dependence of the electrical resistivity is a complex function of temperature, magnetic characteristics and parameters of the electronic spectrum [15, 295]. Above the Curie point, the value of ρ_m does not depend on temperature so that $\rho = \rho_{e-ph} + D \approx AT + D$.

The separation of contributions according to this method was shown in Fig. 261; above T_c the derivative $\partial \rho_{e-ph}/\partial T$ corresponds to $\partial \rho/\partial T$ and is averaged for the β and γ regions.

In the interval from 100 to 1042 K the thermal conductivity of iron has a negative slope. Near the Curie point, $\lambda(T)$ has a minimum [26, 140, 290], the shape of which needs to be somewhat clarified. During the β-γ transformation a slight change in λ is observed, this probably takes place during the γ-δ transformation as well. The uncertainty in the values given at the end of this section is ~2% from 300 to 800 K, 5% above 800 K, and 10% at $T \approx T_c$. Below 700 K, the value of λ_e, calculated as $\lambda_e^g = \lambda - \lambda_g$, becomes smaller than $\lambda_e^L = L_0 T/\rho$ (Fig. 264). Therefore, below 700 K, the Lorenz function of iron becomes smaller than the standard value of L_0 (Fig. 265). Curve 1 corresponds to the ratio of $\lambda - \lambda_g$ to $L_0 T/\rho$. Above 800 K, $L > L_0$, and the maximum value of $\Delta L/L_0 = (L - L_0)/L_0$ does not exceed 15%. This increase in L may be caused by s-d band scattering. Curve 4 shows the results of calculations made by Colquitt [25] for the two-band model. Above 900 K, the experimental values of L do not follow the calculated results and approach L_0; this shows that in the paramagnetic region the electron scattering is elastic.

At temperatures below 700 K, the deviation of L from L_0 is affected by two types of inelastic scattering of s-electrons: 1) the regular electron-phonon scattering, which is most apparent below 300 K, and the electron-electron scattering curve 2), and 2) the scattering of s-electrons by the magnetic inhomogeneities (curve 3) [26].

Figure 266 shows the temperature dependence of the thermoelectric power of iron. It has several extreme and inversion points. Near T_c the thermoelectric power experiences a small jump [296]; considerable jumps are also observed during structural transformations [55].

The kinetic properties of iron are summarized below [39, 140, 291–299].

T, K	100	200	300	400	500	600	700
$\rho \cdot 10^8$, $\Omega \cdot$m	1.2	5.1	10.2	16.4	24.2	33.5	44.8
$a \cdot 10^6$, m^2/s	78.2	30.9	22.7	18.1	14.9	12.4	10.2
λ, W/(m$\cdot$K)	—	93.5	79.0	68.5	60.5	53.5	47.0

T, K	800	900	1000	1200	1400	1600
$\rho \cdot 10^8$, $\Omega \cdot$m	58.6	74.0	91.4	111.9	117	122
$a \cdot 10^6$, m^2/s	8.18	6.30	4.06	6.20	6.60	6.90
λ, W/(m$\cdot$K)	42.0	37.0	32.0	29.0	31.5	34.0

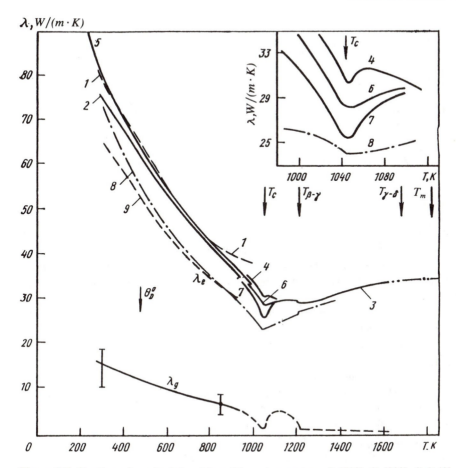

Figure 264 The thermal conductivity of iron (λ) vs. temperature: *1*) [293]; *2*) [298]; *3*) [140]; *4*) [290]; *5*) [48]; *6*) [299]; *7*) [295]; *8*) $\lambda_e^L = L_0 T/\rho$; *9*) $\lambda_e^g = \lambda - \lambda_g$; λ_g is from [23, 140].

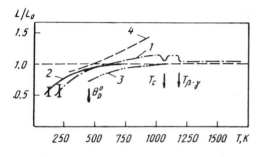

Figure 265 The Lorenz function of iron vs. temperature: *1*) $L/L_0 = \lambda_e^g/\lambda_e^L$; *2*) the calculations of [23]; *3*) the calculations of [295], taking into account the scattering by magnetic inhomogeneities; *4*) the calculations of [25] for a two-band model.

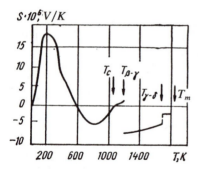

Figure 266 The absolute thermoelectric power of iron (S) as a function of temperature [55].

COBALT

Below 700 K, cobalt has an hcp crystalline structure with the lattice parameters $a = 0.25053$ nm, $c = 0.40892$ nm at room temperature. Above 700 K, cobalt has an fcc structure with the lattice parameter $a = 0.35442$ nm near the transformation point [82]. At temperatures near 1400 K, cobalt transforms from the ferromagnetic into the paramagnetic state.

In the magnetically ordered state caused by a magnetic field, the Fermi surface of cobalt splits and has a different appearance for the electrons with the up-spins and down-spins. Wakoh and Yamashita [41] calculated that the unit cell contains 10.56 electrons with up-spins and 7.44 with down-spins. The Fermi surface of the up-spin electrons is a large sphere at the Γ point in the double-zone scheme. The Fermi surface of the down-spin electrons is shaped in the form of electronic cylinders, parallel to the ML line and connected with each other at the K points. In addition, there are groups of pockets at the L and Γ points. The Fermi surface of paramagnetic cobalt has not been sufficiently investigated. The elastic constants of cobalt ($c_{ij} \cdot 10^{-11}$, Pa) are listed below [38].

T, K	4	73	123	223	298	373
c_{11}	3.195	3.186	3.163	3.109	3.063	3.015
c_{12}	1.661	1.662	1.657	1.655	1.651	1.643
c_{13}	1.021	1.019	1.022	1.021	1.019	1.014
c_{33}	3.736	3.724	3.697	3.631	3.574	3.521
c_{44}	0.824	0.817	0.806	0.777	0.753	0.730
c_{66}	0.767	0.762	0.753	0.727	0.706	0.686

T, K	473	523	573	623	673	708	711
c_{11}	2.946	2.910	—	—	—	—	—
c_{12}	1.628	1.618	—	—	—	—	—
c_{13}	1.009	1.003	—	—	—	—	—
c_{33}	3.440	3.386	3.333	3.301	3.255	3.230	3.236
c_{44}	0.694	0.675	—	—	—	—	—
c_{66}	0.659	0.645	—	—	—	—	—

The only elastic constant which has been measured up to the point of α-β transformation is c_{33}; it undergoes a small jump during the transformation. From

4 to 523 K, all of the elastic constants decrease smoothly with increasing temperature. However, the temperature coefficients of the shear elastic constants are twice as great as those of the compression elastic constants.

Young's modulus of annealed polycrystalline cobalt is equal to 2.09, 1.88 and $1.45 \cdot 10^{11}$ Pa at 293, 673, and 1173 K, respectively [82].

The temperature dependence of the electrical resistivity of cobalt is similar to that of iron (Fig. 267). The α-β transformation near 700 K leads to a considerable hysteresis in the curve of $\rho(T)$. This hysteresis is determined by the setup of the experiment, the rate of heating and the purity of the polycrystalline cobalt sample. Study [92] indicates that, at room temperature, single crystals of α-Co may have a substantial anisotropy in resistivity: $\rho_\perp = 5.544 \ \mu\Omega \cdot cm$, $\rho_\parallel = 10.280$ $\mu\Omega \cdot cm$. Therefore, the data for $\rho(T)$ of polycrystalline α-Co, given in Fig. 267 and at the end of this section, require verification. Most of this data are for quite pure ($r \approx 80$) polycrystalline samples.

Study [140] investigated the properties of single crystals of cobalt and found that a discontinuity exists in the region near T_c. The detailed investigation [300] of the electrical resistivity of polycrystalline cobalt wire showed that only the derivative of resistivity, with respect to temperature, has the anomaly of the λ-type (see Fig. 267). It seems that additional investigations are required in order to clarify the nature of the anomaly near T_c, especially because the anomaly in $\rho(T)$ of iron is not quite the same as that of cobalt (see Fig. 267 and 262).

As can be seen from Fig. 267, the scattering of electrons by magnetic in-

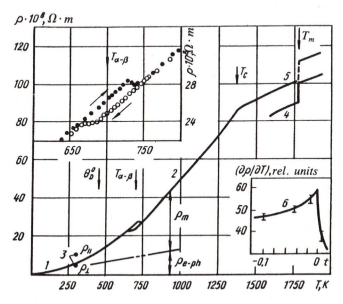

Figure 267 The electrical resistivity of cobalt (ρ) as a function of temperature: *1*) below 300 K [301]; *2*) the average data of [140, 301]; *3*) [92]; *4*) [297]; *5*) [140, 301] plus the data of [297]; *6*) [300].

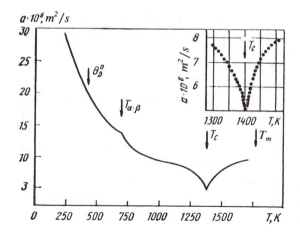

Figure 268 The diffusivity of cobalt (a as a function of temperature [39, 140].

homogeneities substantially exceeds the phonon contribution in cobalt. Study [140] showed that, at low temperatures, the resistance and other kinetic coefficients are better described by the two-band model, which takes into account the s-s and s-d scattering. However, at moderate and especially high ($T > 1000$ K) temperatures, the Kasuya model [15], which takes into account the s-electron scattering by the disordered spins, starts working quite well.

The temperature dependence of the diffusivity of polycrystalline cobalt is shown in Fig. 268 and at the end of this section. The uncertainty in the given values is estimated as 7–13% below 1000 K, and 5% for higher temperatures. Near the Curie point, the curve of $a(T)$ has a minimum described by the equation $a = A t \gamma$, where $\gamma = 0.14$ at $T > T_c$ and $T < T_c$.

On average, the temperature dependence of the thermal conductivity of cobalt (Fig. 269) is similar to that of iron; the uncertainty in the given values is 10%. Figure 269 also shows the estimates of the electronic component of thermal conductivity. Curve *1* was obtained on the basis of the standard W-F-L law at $L =$

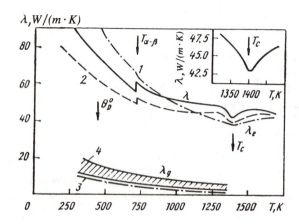

Figure 269 The thermal conductivity of cobalt (λ) as a function of temperature [140, 301]: *1*) λ_e^L; *2*) λ_e^g; *3*) λ_g [301]; *4*) λ_g [140].

L_0. Curve *2* corresponds to $\lambda_e^g = \lambda - \lambda_g$, where λ_g was estimated by the method of [23, 140].

The uncertainty in the estimations of λ_g appears in Fig. 269 as band *4*. The λ_g estimations made in [301] are also shown, they coincide with the lower edge of the band. Although the value of λ_g has a considerable uncertainty, above 700–1000 K the lattice contribution to the total thermal conductivity does not exceed 10%.

As in iron, at temperatures above T_c, the estimates of λ_e by two different methods practically coincide. At moderate and low temperatures, however, they are visibly different (at 500 K the difference reaches 40%). Below 800 K, the value of λ_e^L, obtained from the W-F-L law at $L = L_0$, becomes even greater than λ, which indicates the presence of the inelastic contributions. It is more convenient to discuss their role by analyzing the behavior of the Lorenz function (Fig. 270).

Since cobalt has a rather large value for the thermoelectric power, a correction $L' = L + S^2$ should be introduced for the Lorenz function [140, 301]. The result of applying such a correction to the data is shown by curve *2*. It can be compared with curve *1*, plotted using the original data. Study [140] showed that the difference $\Delta L = L - L_0$ at 500–800 K is due to the inelastic scattering by the magnetic inhomogeneities, and that, at lower temperatures, the inelasticity of the electron-phonon scattering should be taken into account. Above 800 K, $L > L_0$; this can be partially explained by *s-d* band scattering, although this scattering alone produces an increase in L that is too high (curve *4* [301]). More detailed calculations [32, 140] (curves *5* and *6*) showed that the behaviour of $L(T)$ and $\lambda(T)$ can be better described by the one-band model of Kasuya, taking into account the effects of *s-d* band scattering and scattering by spin inhomogeneities.

In cobalt, at moderate and high temperatures, especially above T_c, the electron scattering is elastic, and the Kasuya model [15] closely approximates the scattering by the magnetic inhomogeneities. However, at low temperatures, the band contributions should be taken into account and the two-band model agrees better with the experiment.

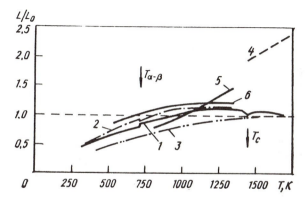

Figure 270 The Lorenz function (L) of cobalt vs. temperature: *1*) $L/L_0 = (\lambda - \lambda_g)/\lambda_e^L$; *2*) $L/L_0 = (L + S^2)/L_0$ [301]; *3*) the inelastic scattering by the spin inhomogeneities are taken into account [140]; *4*) the data of [301]; *5*, *6*) the calculation by the *s-d* model with different parameters of the band structure [25, 140].

Figure 271 shows the temperature dependence of the thermoelectric power of polycrystalline cobalt [55, 301, 302]. The data for α-Co require further investigation; nevertheless, the characteristic temperature dependence for ferromagnetics is readily apparent: a deep minimum at $S(T)$ in the ferromagnetic region. In the paramagnetic region, the value of S is small. Careful experimentation [302] showed that, near the Curie point, the value of the thermoelectric power jumps (see Fig. 271).

The kinetic properties of cobalt are summarized below [140].

T, K	100	200	300	400	500	600	700
$\rho \cdot 10^8$, $\Omega \cdot$m	0.93	3.21	5.99	9.54	14.11	19.87	26.59
$a \cdot 10^6$, m^2/s	—	—	25.5	21.0	17.7	15.5	13.5
λ, W/(m·K)	—	—	87	79	70	63	56

T, K	800	900	1000	1200	1400	1600	1700
$\rho \cdot 10^8$, $\Omega \cdot$m	32.05	40.37	49.56	69.11	87.1	94.8	97.6
$a \cdot 10^6$, m^2/s	11.7	10.5	9.6	8.4	5.3	8.2	7.8
λ, W/(m·K)	56	53	51	49	42	46	—

NICKEL

At standard pressure, nickel has up to the melting point, an fcc structure with the lattice parameter a = 0.35238 nm at 298 K [74]; below 629.63 ± 0.01 K [39, 306] nickel is ferromagnetic. The Fermi surface of ferromagnetic nickel has been established quite well [41]. One of the sheets of the Fermi surface (the electrons with the up-spin) is similar to the Fermi surface of copper and consists of a sphere pulled out in the direction ⟨111⟩ touching the faces of the Brillouin zone; the contact area, however, is considerably smaller than in copper. The Fermi surface of electrons with the down-spin has several sheets, namely two electronic surfaces near point Γ in the fifth and sixth zones and small pockets of holes at point L in the fourth zone and at point X in the third and fourth zones.

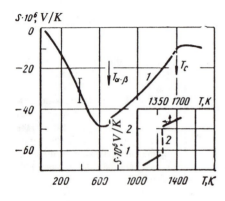

Figure 271 The thermoelectric power (S) of cobalt vs. temperature: *1*) [55]; *2*) [302].

The elastic constants of nickel ($c_{ij} \cdot 10^{-10}$, Pa), given below, were measured from 0 to 760 K in the external magnetic field of 10 kOe [304].

T, K . . .	0	100	200	300	400	500
c_{11}	2,612	2,595	2,553	2,508	2,454	2,396
c_{12}	1,508	1,507	1,501	1,500	1,496	1,490
c_{44}	1,317	1,304	1,270	1,235	1,198	1,159

T, K . . .	600	620	640	700	760
c_{11}	2,322	2,309	2,295	2,261	2,232
c_{12}	1,478	1,479	1,479	1,473	1,464
c_{44}	1,118	1,109	1,100	1,079	1,058

The values of the elastic constants do not really depend on the direction of the external field (within 0.1%). The uncertainty in the absolute values of c_{ij} amounts to $\approx 0.5\%$ [304]. The above results differ from the data obtained using considerably weaker fields by not more than 2%.

The elastic moduli of polycrystalline nickel, which were measured up to 1200 K, are given in Fig. 272 [305].

The temperature dependence of the electrical resistivity of nickel is similar to that of iron and cobalt (Fig. 273). Up to 1700 K, the values were determined with the uncertainty of 1.5–2%.

A well pronounced maximum in the temperature coefficient of the electrical resistivity is observed near the Curie point, the maximum is described by the expressions

$$\partial\rho/\partial T = A' + B' \, t^{-\alpha'} \text{ at } T < T_C$$

$$\partial\rho/\partial T = A + B t^{-\alpha} \text{ at } T > T_C$$

where $\alpha' = -0.3 \pm 0.1$, and $\alpha = 0.0 \pm 0.1$.

The separation of scattering mechanisms is based on the assumption that, above the Curie point, the magnetic contribution is independent of temperature (Fig. 273). It follows from Fig. 273 that the fraction of ρ_m in nickel is smaller than in iron and cobalt. Comparing, at low and moderate temperatures, the one-band model of Kasuya with the two-band s-d model, preference should be given

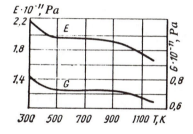

Figure 272 The temperature dependence of Young's modulus (E), and shear modulus (G) of polycrystalline nickel [305].

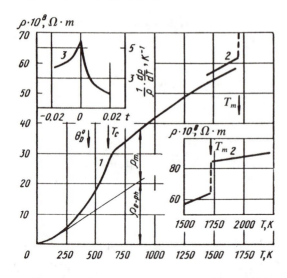

Figure 273 The electrical resistivity of nickel (ρ) vs. temperature: *1*) [306]; 2) [297]; 3) the temperature coefficient of electrical resistivity as a function of the reduced temperature $t = (T - T_c)/T_c$ near the Curie point [306].

to the two-band model. A detailed calculation of the relaxation time and resistivity was made in [303] using the corresponding parameters of the band structure. At 1000 K, the relaxation time for the *s-d* band amounts to 10^{-15} sec, and for the *d*-band 10^{-16} sec. Study [303] also showed that the uncertainty in the Fermi energy of the *s-d* band amounts to $\Delta\varepsilon = 0.6$ eV, and, for the *d*-band, to 6 eV. Such uncertainty makes the fine details of the Fermi surface nonessential for the calculation of the kinetic coefficients; the electronic structure of the 'hot' metal becomes similar to the structure of the liquid metal. Study [303] also showed that, for the resistivity of nickel, the *d*-band could make a substantial contribution to the matrix element of scattering.

The diffusivity coefficient of nickel has a minimum at the Curie point (Fig. 274). The data, given in Fig. 274 and at the end of this section, has an uncertainty of about 10% and refer to a sample with $r = 69 \div 90$ and of about 99.99% purity. Study [309] carefully investigated the diffusivity of nickel near the Curie point and the following expressions were obtained

$$a^{-1} = A'/\gamma' \, [|t|^{-\gamma'} - 1] + B' \quad \text{at} \quad T < T_C$$

$$a^{-1} = A/\gamma \, [|t|^{-\gamma} - 1] + B \quad \text{at} \quad T > T_C$$

where $\gamma' = -0.39 \pm 0.05$, $\gamma = -0.15 \pm 0.05$, $A' = 2.26 \pm 0.20$ s/cm², $B' = 4.55 \pm 0.30$ s/cm², $A = 0.59 \pm 0.10$ s/cm², $B = 6.48 \pm 0.30$ s/cm².

The temperature dependence of the thermal conductivity of nickel is of an extreme nature and has a minimum near the Curie point (Fig. 275). The uncertainty in the data, given at the end of this section, is about 10% [140, 207, 281, 308].

Above 1000 K, the lattice component of thermal conductivity is insignificant, although estimations of it are rather uncertain [140] (as can be seen from the band

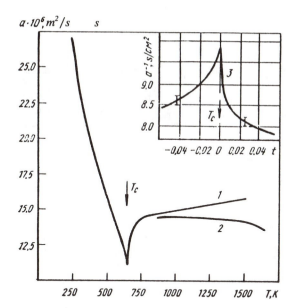

Figure 274 The diffusivity of nickel (a) vs. temperature: *1*) [39]; *2*) [140]; *3*) the reciprocal of the diffusivity vs. the reduced temperature near the Curie point [309].

shown in Fig. 275). The electronic component describes well the behavior of thermal conductivity above 1000 K. It was calculated on the basis of the standard W-F-L law, hence $L \approx L_0$, indicating the dominance of the elastic contributions over electron scattering. However, in the ferromagnetic region, $\lambda_e^L > \lambda$; this can be related to band effects and to the inelastic scattering of electrons by magnetic inhomogeneities.

The role of the inelastic contribution can be analyzed by considering the Lorenz function (Fig. 276). In the paramagnetic region, especially above 1000 K, calculations according to Mott's two-band model [25, 307] result in a considerable

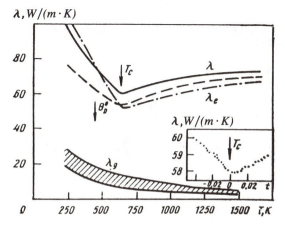

Figure 275 The thermal conductivity of nickel (λ) vs. temperature [140, 281, 308]. In the inset: the thermal conductivity vs. the reduced temperature near the Curie point [309].

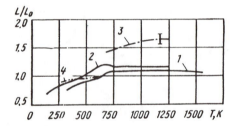

Figure 276 The Lorenz function (L) of nickel vs. temperature: *1*) $L/L_0 = \lambda_e^g/\lambda_e^L$; *2*) $L/L_0 = \lambda/\lambda_e^L$; *3*) calculation based on the *s-d* model [281]; *4*) calculations of electron scattering by magnetic inhomogeneities are taken into account [25, 32].

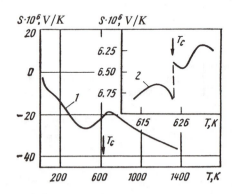

Figure 277 The thermoelectric power (S) of nickel vs. temperature: *1*) [55]; *2*) [302].

overestimation of the Lorenz function as compared with the experimental values. However, below 400–500 K, this model describes well the experimental results (curve *3*). Some inelasticity due to electron scattering by magnetic inhomogeneities [306] occurs for nickel (curve *4*), but its contribution is appreciably smaller than in the case of iron and cobalt.

The temperature dependence of the thermoelectric power of nickel [55, 281] has a minimum in the ferromagnetic region, typical for a classic ferromagnetic metal (Fig. 277). Study [302] observed a jump in the value of the thermoelectric power near the Curie point, this jump is probably related to the critical electron scattering.

The kinetic properties of nickel are summarized below [39, 140, 306].

T, K	100	200	300	400	500	600	700
$\rho \cdot 10^8$, $\Omega \cdot$m	0.98	3.91	7.37	11.91	17.95	26.09	32.02
$a \cdot 10^6$, m^2/s	—	—	22.9	18.7	15.6	12.6	14.2
λ, W/(m·K)	—	—	92	81	72	63	61

T, K	800	900	1000	1200	1400	1600
$\rho \cdot 10^8$, $\Omega \cdot$m	35.51	38.74	41.76	47.29	52.37	—
$a \cdot 10^6$, m^2/s	14.7	14.7	14.8	15.0	15.0	—
λ, W/(m·K)	64	66	68	70	72	73

§12 ACTINIUM, THORIUM, PROTACTINIUM, URANIUM, NEPTUNIUM, PLUTONIUM, AMERICIUM

ACTINIUM

Actinium does not have any stable isotopes; the most long-lived is ^{227}Ac with a half-life of 22 years. Therefore, the physical properties of actinium are poorly known. It has an fcc structure with the melting point at 1323 K and the lattice parameter $a = 0.5311$ nm [320, 321].

THORIUM

At standard pressure and 298 K, thorium has a low-temperature modification, α-Th, with an fcc structure and the lattice parameter $a = 0.50843$ nm. At 1723 K, it has a high-temperature modification, β-Th, with a bcc structure and the lattice parameter $a = 0.411$ nm. The transformation temperature is 1673 ± 25 K [39] or 1638 K [74].

The energy chracteristics of thorium are similar to those of the ordinary transition metals with a broad d-band and s-type electronic conductivity [310]. According to the calculations of Gupta and Loucks [41], the Fermi surface of thorium resembles that of the other fcc transition metals. The main details of the Fermi surface are: a surface of holes around the Γ point which looks like a rounded cube, a dumb-bell-shaped surface of holes centered at the L point along the line ΓL, and an electronic surface on the line ΓK.

The elastic constants of cubic thorium are given below ($c_{ij} \cdot 10^{-10}$, Pa) [311].

T, K	0	40	80	120	160	200	240	280	300
c_{11}	8.103	8.092	8.030	7.970	7.908	7.846	7.787	7.730	7.702
c_{12}	5.031	5.039	5.039	5.045	5.052	5.058	5.069	5.083	5.088
c_{44}	5.046	5.019	4.952	4.880	4.808	4.737	4.663	4.589	4.554

The investigated specimens were 99.96% pure. The values of c_{ij} were calculated by taking into account the thermal expansion of the samples. The given values differ from those obtained earlier (Armstrong, 1959) by not more than 2–3%.

The elastic characteristics of polycrystalline thorium have the following values at $T = 298$ K: $E = 7.38 \cdot 10^{10}$ Pa; $G = 2.81 \cdot 10^{10}$ Pa; $\mu = 0.27$; $\varkappa = 1.64 \cdot 10^{-11}$ 1/Pa [104].

The electrical resistivity of thorium has been thoroughly investigated only for the region below room temperature. According to [312], the resistivity of thorium with $r = 480$ equals to 13.91 $\mu\Omega \cdot$ cm at 273 K. When the value of r decreases to 31, the resistivity increases to 14.52 $\mu\Omega \cdot$ cm. Figure 278 shows the temperature

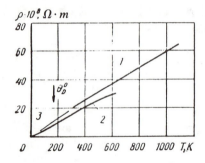

$\rho \cdot 10^9, \Omega \cdot m$

Figure 278 The electrical resistivity of thorium and protactinium (ρ) vs. temperature: *1*) [82] for thorium; *2*) [312, 16] for thorium; *3*) [320] for protactinium.

dependence of the electrical resistivity of thorium. Below 600 K, the nonlinearity in $\rho(T)$ is relatively small, despite the fact that studies [16, 82] found that $\partial^2\rho/\partial T^2 < 0$.

The information on the diffusivity of thorium given below requires further investigation. Its uncertainty is estimated as 20% below 500 K and 30% above 500 K [312]. The thermal conductivity of relatively pure thorium has been studied only for the region below room temperature (Fig. 279). At 200 K, $\lambda = 63$ W/(m·K), this value changes little at higher temperatures. The electronic component of thorium is about 50 W/(m·K), and the difference between the values of λ and λ_e is close to the lattice conductivity for most of the transition metals.

The temperature dependence of the thermoelectric power of thorium has a minimum at moderate temperatures (Fig. 280).

The kinetic properties of thorium are summarized below [39, 48].

T, K	100	200	300	400	500	600	700	800
$a \cdot 10^6$, m^2/s	51.4	41.5	39.2	37.8	36.8	35.9	35.0	34.1
λ, W/(m·K)	48.8	48.8	49.1	49.5	49.8	50.1	50.4	50.8

PROTACTINIUM

Protactinium does not have any stable isotopes, the most long-lived is [231]Pa with the half-life of $3.43 \cdot 10^4$ years. The properties of protactinium are poorly known.

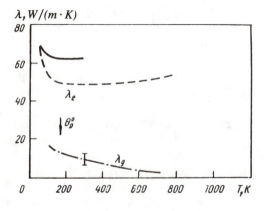

$\lambda, W/(m \cdot K)$

Figure 279 The thermal conductivity of thorium (λ) as a function of temperature [48, 312].

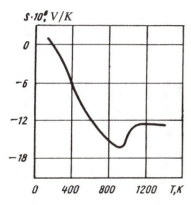

Figure 280 The thermoelectric power of thorium (S) vs. temperature [55].

Protactinium has two modifications: α-Pa with a bct structure and the lattice parameters $a = 0.3929$ nm and $c = 0.3241$ nm at 293 K, and β-Pa with a bcc structure and the lattice parameter $a = 0.381$ nm above 1473 K. The melting temperature of protactinium is 1850 K [321, 322].

At room temperature, the electrical resistivity of protactinium is 19 $\mu\Omega \cdot$ cm, its temperature dependence is similar to that of thorium (see Fig. 278).

URANIUM

At standard pressure, uranium has three polymorphic modifications, the structures and transformation temperatures of which are given in Table 6. At 23 K and 37 K, isostructural phase transitions of the first order are observed in orthorhombic $\alpha - U$. At 42 K, its physical properties have anomalies which indicate that a second order like transition is taking place [74]. Near ~1K uranium transforms into the superconducting state [104]. Uranium, contrary to thorium, has a narrow $5f$-electron band in addition to the s- and d-bands (energy levels).

Table 47 lists the elastic constants of orthorhombic uranium for the temperature interval from 44 to 923 K. Figure 281 shows the temperature dependences of the elastic moduli and volume compressibility for polycrystalline α-U. The investigated sample was 99.98% pure; the uncertainty in the values of the moduli is $\approx 0.2\%$. Significant changes in the values of E, G and $\varkappa$ at 37 K show that a softening of the crystal lattice takes place during the isostructural phase transformation [314].

Figure 282 shows the temperature dependence of the electrical resistivity of uranium [96, 104]. The distinguishing features are a negative curvature and a decrease in the electrical resistivity during the α-β and β-γ transformations. Figure 282 also shows the electrical resistivity of single crystals of α-U at 273 K. The electrical resistivity of polycrystalline α-U depends substantially on the thermomechanical history of the sample because of its significant anisotropy. At room temperature, the resistivity values can range between 25 and 40 $\mu\Omega \cdot$ cm, which

Table 47 The elastic constants of uranium ($c_{ij} \cdot 10^{-11}$, Pa) [313]

T, K	c_{11}	c_{55}	c_{12}	c_{66}	c_{13}	c_{22}	c_{23}	c_{33}	c_{44}
44	1.500	0.892	0.275	0.849	0.345	2.085	1.123	2.868	1.407
98	2.063	0.879	0.425	0.833	0.226	2.081	0.098	2.856	1.384
173	2.138	0.826	0.441	0.799	0.216	2.046	1.091	2.791	1.332
248	2.151	0.771	0.454	0.766	0.216	2.011	1.081	2.721	1.280
298	2.148	0.734	0.465	0.743	0.218	1.986	1.076	2.671	1.244
373	2.132	0.677	0.479	0.708	0.223	1.947	1.069	2.597	1.193
473	2.097	0.601	0.494	0.661	0.235	1.889	1.044	2.485	1.122
573	2.049	0.513	0.517	0.607	0.257	1.827	1.009	2.369	1.045
673	1.984	0.426	0.537	0.546	0.282	1.760	0.987	2.249	0.958
773	1.904	0.340	0.566	0.477	0.312	1.681	0.971	2.118	0.873
873	1.804	0.253	0.612	0.388	0.349	1.584	0.959	1.977	0.780
923	1.742	0.211	0.630	0.344	0.374	1.535	0.953	1.907	0.734

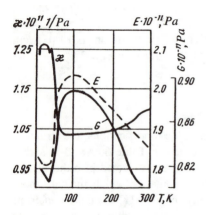

Figure 281 Young's modulus (E), shear modulus (G) and volume compressibility ($\varkappa$) of α-U vs. temperature [314].

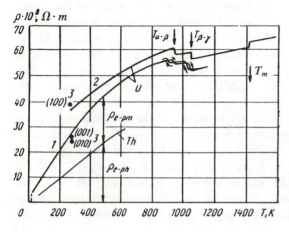

Figure 282 The electrical resistivity of uranium (ρ) vs. temperature: *1*) [104]; *2*) [315]; *3*) [104], for single crystals.

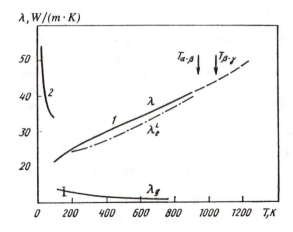

Figure 283 The thermal conductivity of uranium (λ) vs. temperature: *1*) [48]; 2) [322].

are the typical values for the principal directions. The actual observed values for samples, which have undergone different treatments, were within this range [104].

Figure 282 also shows the temperature dependence of the electrical resistivity of thorium. As a first approximation, it is identical to the electron-phonon component of the resistivity of uranium caused by the ordinary scattering of the s- and d-carriers. In this case, the difference $\rho_U - \rho_{Th} \approx \rho_{e-pm}$ is probably due to an additional scattering on the $5f$-levels, and may be caused by a paramagnon scattering mechanism, according to calculations in [16] for neptunium and plutonium.

The thermal conductivity of uranium (Fig. 283) has not been sufficiently investigated; it increases with temperature, and $\lambda_e^L = \lambda$ within the uncertainty in the values of λ ($\approx 20\%$).

Figure 284 shows the absolute thermoelectric power of uranium [55]. Two facts should be pointed out: the increase in the values of S, and small changes during structural transformations.

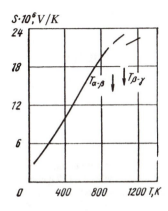

Figure 284 The absolute thermoelectric power of uranium (S) vs. temperature [55].

NEPTUNIUM

At standard pressure, neptunium has three polymorphic modifications [74]. The orthorhombic modification, α-Np, has the lattice parameters $a = 0.4723$ nm, $b = 0.4887$ nm, $c = 0.6663$ nm at 293 K. At 554 K, this modification transforms into a tetragonal modification, β-Np, with the lattice parameters $a = 0.4897$ nm, $c = 0.3388$ nm at 586 K. At 850 K, the tetragonal structure, β-Np, transforms into a bcc structure with $a = 0.352$ nm at 873 K. The melting temperature of neptunium is $T_m = 913.2$ K [74].

Similar to uranium, neptunium has the s- and d-like bands and an additional narrow band of $5f$-electrons which is probably located near the Fermi energy of the principal carriers.

The elastic properties of α-Np are similar to those of α-U and differ considerably from those of plutonium. For the very pure α-Np, obtained by an electrolytical refining, the elastic moduli have the following values at 298 K: $E = 1.85 \cdot 10^{11}$ Pa; $G = 0.76 \cdot 10^{11}$ Pa; $K = 1.12 \cdot 10^{11}$ Pa. These moduli increase with decreasing temperature, and, at 77 K, they have the following values: $E = 2.185 \cdot 10^{11}$ Pa; $G = 0.895 \cdot 10^{11}$ Pa; $K = 1.294 \cdot 10^{11}$ Pa [315].

The temperature dependence of the electrical resistivity of α-Np has an appearance which is unusual for nonmagnetic metals, it becomes saturated near 300 K (Fig. 285); the uncertainty in the determination of the absolute values is $\approx 10\%$.

Study [116] applied the model of paramagnon scattering to the analysis of the electrical conductivity of neptunium. Figure 285 (curve 2) shows the results of these calculations. The following parameters were used: the Fermi temperature of a 'narrow' band, created by large spin fluctuations (paramagnons), $T_F = 750$ K; the resistivity $\rho_\infty = 71.5$ $\mu\Omega \cdot$ cm at $T \to \infty$; the Stoner factor $S_{St} = 10$; $\xi = k_{F_s}/k_{F_i} = 0.5$, where k_{F_s} and k_{F_i} are the Fermi impulses of the broad s- and narrow i-bands. The resistivity of thorium was assumed to be equivalent to the electron-phonon contribution. The total resistivity $\rho = \rho_{e-ph} + \rho_{e-pm}$ describes well the experimental values.

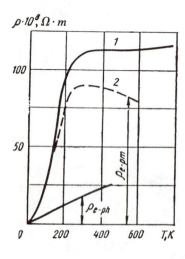

$\rho \cdot 10^{9}, \Omega \cdot m$

Figure 285 The electrical resistivity of neptunium (ρ) vs. temperature [16]: 1) the total resistivity; 2) the calculations of [16] for the electron-paramagnon contribution.

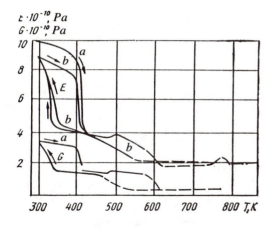

$t \cdot 10^{-10}$ Pa
$G \cdot 10^{-10}$ Pa

Figure 286 Young's (E) and shear (G) moduli of plutonium vs. temperature [317].

Other kinetic properties of neptunium were investigated insufficiently. Study [319] lists the values for several constants at 300 K. The Lorenz number substantially exceeds the standard value. This can be explained by the inelastic contributions and, mainly, by the greater proportion of the lattice contribution insofar as the total thermal conductivity of neptunium is small. It is expected that thermal conductivity should increase with temperature because its electronic component $\lambda_e^L = L_0 T/\rho$ increases, while ρ is almost independent from temperature.

PLUTONIUM

At standard pressure, plutonium has six crystalline modifications. The crystallographic parameters and transformation temperatures of these structures are given in Table 6.

The main distinguishing feature of the electronic states of plutonium is the presence of a narrow band of 5f-electrons. It is located in the vicinity (in comparison to the width of the thermal band) of the Fermi level of the s- and d-electrons which determine the conductivity of this metal [310].

The temperature dependences of the elastic moduli, given in Fig. 286, were determined by the dynamic method [317]. Curve (a) was obtained by heating a sample up to 470 K and then cooling it down to room temperature. The unusual features are: the curvature of $E(T)$ in the α-phase region and the high temperature of $T_{\alpha-\beta} = 410$ K. Similar behavior, however, was observed during the first heating of other specimens. Curve (b) was obtained during the second heating of the samples up to 850 K; the curvature of $E(T)$ in the region of the α-phase is small, $T_{\alpha-\beta} = 395$ K.

The curve of $G(T)$ was obtained during the fourth heating. At $T_{\alpha-\beta}$, both moduli decrease by about 50%. The transformation at $T_{\beta-\alpha} = 483$ K is characterized by a 0.1% increase in the values of the moduli despite a 3.5% decrease in density. Unreliable segments of the curves (strong sound fading) are shown in Fig. 286 by the broken lines.

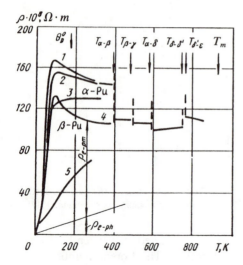

$\rho \cdot 10^{\circ}_{\circ}, \Omega \cdot m$

Figure 287 The electrical resistivities of plutonium and americium (ρ) vs. temperature: *1*) the axis of the sample is parallel to the axis [010] [16]; *2*) for a polycrystal [104]; *3*) the axis of the sample is perpendicular to the axis [010] [16]; *4*) for a polycrystal [104]; *5*) the data for americium (321, 322).

The volume compressibility of α-Pu at 300 K is $\varkappa = (2.1 \div 2.5) \cdot 10^{-11}$ 1/Pa, and for β-Pu at 473 K $X = (2.3 \pm 1) \cdot 10^{-11}$ 1/Pa for the pressure interval from 0 to 2 kbar [317].

The elastic constants of δ-Pu are also known [318]. The addition of 1% of Ga (by mass) stabilizes δ-Pu at room temperature. This material, which has the density of 15.75 ± 0.01 g/cm^3 at $T = 300$ K, has the following elastic constants ($c_{ij} \cdot 10^{-10}$, Pa); $c_{11} = 3.628 \pm 0.036$; $c_{12} = 2.673 \pm 0.027$; $c_{44} = 3.359 \pm 0.01$; $K = 2.991 \pm 0.030$. High values of the anisotropy are typical for fcc plutonium: $A = 7.03$.

Figure 287 shows the temperature dependence of the electrical resistivity of plutonium. Below 300 K, the data refer to the single crystals of α-Pu with the axis directions parallel and perpendicular to the direction [010] [16], as well as to the quenched modification of β-Pu.

Plutonium has a low thermal conductivity and $\lambda_e \approx \lambda_g$ in the neighborhood of 100 K (Fig. 288). Above 200 K, the thermal conductivity of plutonium has a considerable positive temperature coefficient. Study [323] gives the data on the electrical resistivity, diffusivity and thermal conductivity of ~99.99% pure polycrystalline plutonium for the temperature range from room temperature to melting point. The kinetic properties are discontinuous at the points of structural transformations. These kinetic properties are given below [323].

	α phase		β phase		γ phase		δ phase		ϵ phase	
T, K	298	373	423	473	498	548	598	998	773	823
$\rho \cdot 10^8$, $\Omega \cdot m$ . . .	142	140	108	108	107	107	100	100	114	114
$a \cdot 10^6$, m^2/s . . .	1.78	2.10	3.10	3.31	3.53	3.96	4.40	4.85	5.04	5.25
λ, W/(m·K). . .	5.20	6.62	7.87	8.67	8.97	10.46	10.97	12.10	12.1	12.6
$L \cdot 10^8$, V^2/K^2. . .	2.48	2.48	1.99	1.98	1.93	2.04	1.83	1.73	1.78	1.75

The nature of the anomalous behaviour of the electrical resistivity and thermal conductivity of actinides remains unclear. Study [16] proposed that they are metals with strong exchange interactions, and that their physical properties can be explained by the paramagnon model. In the high-temperature approximation, the electrical resistivity of plutonium is described by equation (19) with the following parameters $\rho_\infty = 95$ $\mu\Omega \cdot$ cm, $S_{St} = 10$, $T_{F_i} = 280$ K, $\xi = 0.37$; 0.4 and 0.53 for the curves *1*, *2* and *3*, respectively (Fig. 287). As in neptunium, the regular electron-phonon component of plutonium was assumed to be equal to the resistivity of thorium.

Equation (20) described the behavior of thermal conductivity and the Lorenz function according to this model. According to equation (20), if the paramagnon mechanism of electron scattering is realized, the Lorenz number should decrease relative to the standard value of L_0. Study [323] showed that some decrease in L actually happens, especially in the high temperature phases. These results can be used to further develop the concept of the paramagnon scattering mechanism in actinides. It is not quite clear why the Lorenz number of the α-phase is close to L_0. The α-region, however, has the most pronounced nonlinearity in the behavior of $\rho(T)$, which is typical for paramagnon scattering. A maximum of $\rho(T)$ can be observed in Fig. 287.

Study [310] proposed an alternative model. In actinides the complex crystal structure creates optical branches of the phonon spectrum. This, in turn, gives rise to a strong electron scattering, so that two temperature regions can be established for plutonium. According to the data of [310], below 100 K, plutonium is an ordinary metal with a positive temperature coefficient of resistivity. However, strong scattering by optical phonons causes such an increase in electrical resistivity that, above 100 K, the mean free path of electrons approaches the interatomic distance. In this case, the character of the conductivity of plutonium resembles that of an amorphous metal with a typically 'jumpy' conductivity. The resistivity in the region of the negative temperature coefficient is described by the expression: $\rho = B_{exp}(-W/T)$ where, for polycrystalline α-Pu, $B = 140$ $\mu\Omega \cdot$ cm, $W = 19.8$ K and, for β-Pu, $B = 106$ $\mu\Omega \cdot$ cm, $W = 9.0$ K. Unfortunately, study [310] did not give the values for other kinetic properties calculated from this model.

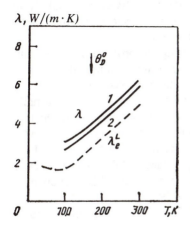

Figure 288 The thermal conductivity of single crystal α-Pu (λ) vs. temperature [320]: *1*) the axis of the sample is parallel to the direction [020]; *2*) the same for the perpendicular direction.

Therefore, the question of the dominant mechanism for high temperature electron scattering in plutonium remains open.

Study [321] mentioned that the thermoelectric power of the actinides is comparable to that of the rare earth metals.

The Hall coefficient of plutonium has a complex temperature dependence [321]. For the β- and α-phases its sign changes from negative to positive near 20 and 200 K, respectively. Above 300 K, its temperature dependence becomes weaker. For δ-Pu this coefficient is positive and decreases rapidly, approaching, at 200 K, the typical values of the α- and β-phases.

AMERICIUM

The most long-lived isotope of americium is [242]Am, which has a half-life of 100 years. Below 1350 K, americium has a hexagonal structure with the lattice parameters $a = 0.3480$ nm and $c = 1.1240$ nm. In the interval of 1350–1447 K, (i.e. up to the melting point) it has an fcc structure with $a = 0.4894$ nm [321, 322].

At room temperature, the electrical resistivity of americium is 68 $\mu\Omega \cdot$ cm; its temperature dependence is shown in Fig. 287 [321; 322].

In conclusion, the behavior of the kinetic properties of metals at high temperatures is now quite well established. This behavior has been most completely studied for nontransition metals, although for many of them the existing data require verification and additional measurements with purer single crystals. For these metals, the thermal and thermogalvanomagnetic characteristics have been studied insufficiently, especially near the melting point and in the liquid state. The typical features of the kinetic properties of nontransition metals are positive values of the temperature coefficient of resistivity and positive curvature of $\rho(T)$ for the solid and liquid states, as well as a negative coefficient of thermal conductivity above 100–150 K. The thermal energy is transported mainly by electrons, as a result of which the lattice component is difficult to detect because of experimental uncertainty. Above 200 K, the Lorenz number usually does not differ considerably from the standard value of L_0. It should be mentioned that the anisotropy in the kinetic properties of these metals remains up to the melting point.

The common concepts in the behavior of the kinetic properties of the transition metals should now be pointed out. Above room temperature, the temperature coefficient of resistivity decreases for most of them, as thermal conductivity simultaneously increases. The lattice component of thermal conductivity is usually small for these metals. In some cases, however, (for example manganese, the rare earth metals, and actinides), at moderate temperatures, it can be close in magnitude to the electronic component. It should be mentioned that even for 'bad' transition metals with high electrical resistivity high values of thermoelectric power are not usually observed (the exception is probably europium). This is in contrast to the nontransition semimetals.

The kinetic properties of the transition metals near the melting point and in the liquid state are the least known. A recent development in high-temperature experimental technique overcomes many problems. Nonstationary methods of measurements permit the obtaining of physical properties for a broad temperature interval in a split-second. The implementation of impulse heating helped obtain the electrical resistivity of liquid tungsten, molybdenum and other refractory metals and to study the influence of high pressures (up to 50 kbar) [324, 325].

For the diffusivity measurements of metals near the melting point a method of dynamic planar temperature waves was developed by the present author together with the postgraduate students S. A. Ilinykh, S. G. Taluts, and V. S. Polevoy [326–328]. For a number of aggressive and refractory metals, the diffusivity data for the solid and liquid states were obtained.

Figure 263 shows, in the inset, the results of the diffusivity measurements of pure iron (Johnson and Mathew); the metal impurities included in % by mass: Si—0.001, Ni—0.0002, Cu, Mg, Na, and Ag—0.0001; the relative residual resistance was $r = 128$. The measurements were made with the rates of heating of 180 and 350 K/s. The average results are given below.[1]

T, K	1500	1550	1600	1650	1700	1750	1800	1850
$a \cdot 10^6$, m^2/s	6.9	7.0	7.2	7.5	6.1	6.2	7.0	7.0

This method was also used for investigating the high temperature diffusivity of titanium in the solid and liquid states. Samples of 99.99% pure polycrystalline titanium were used. The measurements were made with rates of heating of 500, 600 and 1000 K/s. The results of these measurements are given below.[2]

T, K	1500	1600	1700	1800	1900	1950	1980	2000
$a \cdot 10^6$, m^2/s	8.7	8.9	9.0	9.0	9.0	9.1	10.0	10.0

This method was also used for investigating the temperature dependence of diffusivity in single crystals of molybdenum with a relative residual resistance of $r = 1370$ in the direction perpendicular to [110].

The following results were obtained.[3]

T, K	2200	2400	2600	2800	2850	2900	3000
$a \cdot 10^6$, m^2/s	24	22	20	17.5	16.2	12	12

Although these data are of a preliminary nature, it should nevertheless be pointed out that the diffusivity jump during melting is appreciably greater in molybdenum than in iron or titanium and is, also, of an opposite sign.

An interesting feature distinguishing the behavior of the kinetic properties of the transition and nontransition metals is the relatively small increase in the kinetic

[1]1500–1750 K) the solid state; phases: α (1500–1650 K) and δ (1700 and 1750 K); 1800 and 1850 K) the liquid state.
[2]1500–1950 K) the solid state; β-phase; 1980 and 2000 K) the liquid state.
[3]2200–2850 K) the solid state; 2900–3000 K) the liquid state.

properties of transition metals during melting and the disappearance of the anisotropy in conductivity near the melting point.

Studies [8, 140, 326] explained such behavior by a decrease in the mean free path of electrons down to the interatomic distance. For small mean free paths, equation (6), which is a consequence of the Heizenberg uncertainty principle, should be taken into account. In this case, the uncertainty in the Fermi wave number $\delta_{k_F} = \Lambda^{-1}$ should lead to the uncertainty in its anisotropy $\delta\alpha_F = \delta k_F/k_{F_{av}} = (\Lambda \cdot k_{F_{av}})^{-1}$, and consequently, to its decrease in comparison with the 'ideal' value of $\alpha_F^0 = \Delta k_F^0/k_{F_{av}}$. If $\alpha_F^0 = \Delta_{k_F}/k_{F_{av}} = (k_{F_\perp} - k_{F_\parallel})/k_{F_{av}}$ is the anisotropy of the Fermi wave number for a hexagonal crystal, then $\alpha_F^0 \approx (S_\parallel - S_\perp)/S_{F_{av}} = \Delta S_{F_{av}}/S_{F_{av}}$, where $S_\parallel$ and $S_\perp$ are the projection areas of parts of the Fermi surface in the corresponding directions of the reprocical space. Then, the influence of the uncertainty principle leads to the following value of the anisotropy for a given mean free path $\alpha_F = \alpha_F^0 - \delta\alpha_F = a_F^0 - (\Lambda k_{F_{av}})^{-1}$. It follows from equation (9) that, at high temperatures, the anisotropy in resistance is determined by the ratio of projections of the corresponding parts of the Fermi surface only. If to take into account the uncertainty principle, the effective anisotropy will be less than the 'ideal' value of α_ρ^0: $\alpha_\rho \approx \alpha_\rho^0 - \delta\alpha_F = \alpha_F^0 - \delta\alpha_F$. Thus, the decrease in the mean free path of electrons leads to a decrease in the anisotropy of the electrical and thermal conductivities and diffusivity; the anisotropy of a number of metals disappears long before the melting point. Since the mean free path of electrons is determined from the value of electrical resistivity $\Lambda \sim \rho^{-1}$, a correlation can be established between the value of electrical resistivity and its anisotropy $\alpha_\rho = \alpha_\rho^0 - A\rho$, where A is a constant. This correlation really takes place [8], although the degree of its applicability is not clear.

Similar reasoning may be applied to the jump in conductivity at melting. Thus, if the mean free path is short, it does not change considerably during melting, and the magnitude of the jump is determined only by the change in the Fermi characteristics of the electrons [326]. The increase in the uncertainty of these characteristics decreases the magnitude of the jump during melting [8, 140, 326].

Generally speaking, the physics of the high-temperature kinetic properties of metals are not sufficiently developed. The 'classic' extrapolations of the low- and moderate-temperature relationships (for example, $\rho \sim T$, $\Lambda \sim T^{-1}$) to the region of high temperatures are often unacceptable, as can be seen from the examples given here. One important path for the development of high-temperature physics of metals would be an improvement in the experimental methods with the purpose of refining the existing data and, in some cases, of obtaining new data (for example, the thermogalvanomagnetic properties). The other path would be a creation of a consistent theory of kinetic properties of metals at high temperatures, for which the necessary prerequisites are already available.

REFERENCES

1. Madelung, O. Introduction to solid state theory. New York, Springer-Verlag, 1978.
2. Ziman, J. M. Electrons and phonons. London, Oxford University, 1963.
3. Bass, J. Adv. Phys., 1972, Vol. 21, pp. 433–604.
4. Novikov, I. I. In: Review of thermophysical properties of substances. Timely problems of thermodynamics. Moscow, Inst. Vys. Temp. AN SSSR, 1977, No. 3, pp. 3–42.
5. Pervakov, V. A. Metallophysics. Kiev, Naukova dumka, 1970, No. 30, pp. 5–16.
6. Kraftmakher, Ya. A. J. Sci. Eng. Res., 1973, Vol. E32, pp. 626–632.
7. Startsev, V. E., Dyakina, V. P., Cherepanov, V. I., et al. Zh. Eksp. Teor. Fiz., 1980, Vol. 79, pp. 1335–1344.
8. Zinov'ev, V. E. Fiz. Tverd. Tela, 1978, Vol. 20, pp. 2249–2253.
9. Mott, N. F. Proc. Roy. Soc., 1936, Vol. A153, pp. 699–716.
10. Wilson, A. H. Proc. Roy. Soc., 1938, Vol. A167, pp. 580–593.
11. Aisaka, T. and Shimizu, M. J. Phys. Soc. Japan, 1970, Vol. 28, pp. 646–654.
12. Smirnov, I. A. and Tamarchenko, V. I. Electronic thermal conductivity in metals and semi-conductors. Leningrad, Nauka, 1977, 151 pp.
13. Appel, J. Phil. Mag., 1963, Vol. 89, pp. 1071–1073.
14. Volkenstein, M. V., Dyakina, V. P. and Startsev, V. E. Phys. Stat. Sol., 1978, Vol. 57b, pp. 9–48.
15. Vonsovskii, S. V. Magnetism. Moscow, Nauka, 1971, 1032 pp.
16. Jullien, R., Béal-Monod, M. T., and Coqblin, B. Phys. Rev., 1974, Vol. 9, pp. 1441–1457.
17. Béal-Monod, M. T. and Mills, D. L. Sol. Stat. Comm., 1973, Vol. 13, pp. 1707–1711.
18. Kasuja, T. and Kondo, A. Sol. Stat. Comm., 1974, Vol. 14, pp. 253–256.
19. Richard, T. G. and Gel'dart, D. J. W. Phys. Rev. Lett., 1973, Vol. 30, pp. 290–294.
20. Su, D. J. Low Temp. Phys., 1976, Vol. 24, pp. 701–708.
21. Alexander, S., Helman, J. S., and Balberg, I. Phys. Rev., 1976, Vol. 130, pp. 304–315.
22. Patashinskii, A. Z. and Pokrovskii, V. L. Fluctuation theory of phase transitions. Moscow, Nauka, 1975, 255 pp.
23. Williams, R. K. and Fulkerson, W. In: Thermal Conductivity. Proc. First Conf., W. Laf, USA, 1968, pp. 1–120.
24. Oskotskii, V. S. and Smirnov, I. A. Defects in crystals and thermal conductivity. Leningrad, Nauka, 1972, 160 pp.
25. Colquitt, L. Phys. Rev., 1965, Vol. 139, pp. A1857–1859.

26. Zinov'yev, V. E., Abel'skii, Sh. Sh., Sandakova, M. I., et al. Zh. Eksp. Teor. Fiz., 1974, Vol. 66, pp. 354–360.

27. Stanley, H. E. Introduction to phase transitions and critical phenomena. New York, Oxford University Press, 1971.

28. Ma, S. K. Modern theory of critical phenomena. New York, Benjamin, 1976.

29. Kalashnikov, V. P., Auslender, M. I., and Karyagin, M. I. Fiz. Metal Metalloved, 1980, Vol. 50, pp. 688–695.

30. Stern, H. J. Phys. Chem. Sol., Vol. 36, pp. 153–161.

31. Lykov, A. V. Theory of thermal conductivity. Moscow, Vyssh. shkola, 1967, 599 pp.

32. Dik, E. G. and Abel'skii, Sh. Sh. Fiz. Metal Metalloved, 1974, Vol. 37, pp. 1305–1308.

33. Dik, E. G., Volkov, V. N., Abel'skii, Sh. Sh. et al. Zh. Eksp. Teor. Fiz., 1975, Vol. 69, pp. 611–616.

34. Leibfried, G. Microscopic theory of the mechanical and thermal properties of crystals. Leningrad: Fizmatgiz, 1963, 312 pp.

35. Alers, G. A. and Neighbours, J. R. J. Appl. Phys., 1957, p. 1514.

36. Mason, W. P., ed., Physical acoustics: principles and methods, Vol. 3, Pt. B: Lattice dynamics. New York, Academy Press, 1969.

37. Truell, P., Elbaum, C., and Chick, B. B. Ultrasonic methods in solid state physics. New York, Academy Press, 1969.

38. Fisher, E. S. and Dever, D. Trans. Metal. Soc. AIME, Vol. 239, pp. 48–57.

39. Touloukian, Y. S., ed. Thermal properties of matter. Vol. 10. Thermal diffusivity. New York, Plenum, 1973, 649 pp.

40. Chi, T. C. J. Phys. Chem. Ref. Data, 1979, Vol. 8, pp. 339–438.

41. Cracknell, A. P. and Wong, K. C. The Fermi surface. Oxford, Clarendon Press, 1973.

42. Stowinski, T. and Trivisonno, J. Phys. Chem. Sol., 1969, Vol. 30, pp. 1276–1279.

43. Martinson, R. H. Phys. Rev., 1969, Vol. 178, pp. 902–911.

44. Fritsch, G., Geipel, F., and Prasetyo, A. J. Phys. Chem. Sol., 1973, Vol. 34, pp. 1961–1969.

45. Marquardt, W. R. and Trivisonno, J. J. Phys. Chem. Sol., 1965, Vol. 26, pp. 273–278.

46. Fritsch, G. and Bube, H. Phys. Stat. Sol., 1975, Vol. 30a, pp. 571–576.

47. Zhernov, A. P. and Kagan, Yu. M. Fiz. Tverd. Tela, 1978, Vol. 20, pp. 3306–3320.

48. TPRC data book. Series on thermal properties. Vols. 1–3, Thermal conductivity. New York: Plenum publishing corp., 1969, 421 pp.

49. Shpil'rain, E. E., Yakimovich, K. A., Totskii, E. E., et al. In: Thermophysical properties of alkali metals. Kirillin, V. A., ed. Moscow, Standartov publishing house, 1970, 487 pp.

50. Zhernov, A. P. Fiz. Tverd. Tela, 1980, Vol. 22, pp. 575–580.

51. Filippov, L. P. Investigation of thermal conductivity of liquids. Moscow, MGU publishing house, 1970, 239 pp.

52. Langeneine, J. and Mayer, H. Z. fur Phys, 1972, B249, pp. 386–399.

53. Cook, J. G. and Laubitz, M. J. Therm. Conductivity, Vol. 14, New York, 1976, pp. 105–111.

54. Sharma, P. K., Prakash, J., and Menrotra, K. N. Acta Phys. Pol., 1979, Vol. A56, pp. 755–764.

55. Vedernikov, M. V. Adv. in Phys., 1969, Vol. 18, pp. 337–370; Vedernikov, M. V. and Burkov, A. T. In: Thermoelectricity in metallic conductors. New York: Plenum Press, 1973, pp. 77–90.

56. Gutman, E. and Trivisonno, J. Phys. Chem. Sol., 1967, Vol. 28, pp. 805–809.

57. Kollarits, E. and Trivisonno, J. J. Phys. Chem. Sol., 1968, Vol. 29, pp. 2133–2139.

58. Hufner, S., Wertheim, W., and Wernick, J. H. Phys. Rev., 1973, Vol. 88, pp. 4511–4524.

59. Eastman, D. E. J. Appl. Phys., 1969, Vol. 40, pp. 1387–1394.

60. Bansil, A., Schwartz, L., and Ehrenreich, H. Phys. Rev. 1975, Vol. B12, pp. 2893–2907.

61. Ramakanth, A., Chaterjee, A., and Sinha, S. K. Indian J. Phys., 1975, Vol. 49, pp. 373–384.

62. Chang, Y. A. and Himmel, L. J. Appl. Phys., 1966, Vol. 37, pp. 3567–3572.

63. Matula, R. A. J. Phys. Chem. Ref. Data, 1979, Vol. 8, pp. 1147–1298.

64. Moriarty, J. H. Phys. Rev., 1970, Vol. B1, pp. 1363–1370.
65. Dagents, L. J. Phys., 1976, Vol. F6, pp. 1801–1819.
66. Uspenskii, Yu. A., Mazin, I. I., and Savitskii, E. M. Fiz. Metal Metalloved, 1980, Vol. 50, pp. 231–241.
67. Yamashita, J. and Asano, S. Progr. Theor. Phys., 1973, Vol. 50, pp. 1110–1119.
68. Busch, G. and Güntherhodt, H. J. Phys. Kondens. Mater., 1967, Vol. 6, pp. 325–362.
69. Blatt, F. J., Schroeder, P. A., Foiles, K. A., and Greig, D. Thermoelectric power of metals. New York, Plenum Press, 1976.
70. Schimizu, M., Takahashi, T., Katsuki, A. J. Phys. Soc. Japan, 1963, Vol. 18, pp. 240–248.
71. Christensen, N. E. Phys. Stat. Sol., 1972, Vol. 54b, pp. 551–563.
72. Neighfour, S. and Alers, G. A. Phys. Rev., 1959, Vol. 111, pp. 707–712.
73. Narayana, K. L. and Swamy, K. M. Mater Sci. Eng., 1975, Vol. 18, pp. 157–158.
74. Tonkov, E. Yu. Phase diagrams of the elements at high pressures. Moscow, Nauka, 1979, 192 pp.
75. Rowland, W. D. and White, J. S. J. Phys. F: Met. Phys., 1972, Vol. 2, pp. 231–235.
76. Mitchell, M. H., J. Appl. Phys., 1975, Vol. 46, pp. 4742–4746.
77. Chi, T. C. Phys. Chem. Ref. Data, 1979, Vol. 8, pp. 439–497.
78. Slutsky, L. J. and Garland, C. W. Phys. Rev., 1957, Vol. 107, pp. 972–976.
79. Alderson, J. E. A. and Hurd, C. M. Phys. Rev., 1975, Vol. B12, pp. 501–510.
80. Melidis, K. Phys. Stat. Sol., 1978, Vol. A47, pp. K27–K30.
81. Blatt, F. J. Helv. Phys. Acta, 1968, Vol. 41, pp. 693–700.
82. Samsonov, G. V., ed. Properties of elements, Pt. 1. Moscow, Metallurgiya, 1976, 599 pp.
83. Cook, J. V. and Van der Meer, M.-P. Phys. F: Met. Phys., 1973, Vol. 3, pp. L130–L133.
84. Rottman, C. and Van Zytveld, J. B. J. Phys. F: Met. Phys., 1979, Vol. 9, pp. 2049–2056.
85. Koster, W. and Franz, H. Metal Rev., 1961, Vol. 6, pp. 1–55.
86. Rashid, M. S. and Kayser, F. Y. J. Less. Comm. Met., 1971, Vol. 25, p. 107.
87. Voronov, F. F. and Stal'gorova, O. V. Zh. Eksp. Teor. Fiz., 1965, Vol. 49, pp. 755–759.
88. Güherodt, H. J., Hauser, E., Kunzi, H. U., Evers, J., and Klids, E. K. J. Phys. F: Met. Phys., 1976, Vol. 6, pp. 1513–1522.
89. Alers, G. A. and Neighbours, J. R. J. Phys. Chem. Sol., 1958, Vol. 7, pp. 58–64.
90. Garland, C. W. and Silverman, J. Phys. Rev., 1960, Vol. 119, pp. 1218–1222.
91. Chang, Y. A. and Himmel, L. J. Appl. Phys., 1966, Vol. 37, pp. 3787–3790.
92. Hurd, C. M. Adv. in Phys., 1974, Vol. 23, pp. 315–433.
93. Filippov, L. P. Measurements of thermal properties of solid and liquid metals at high temperatures. Moscow, MGU publishing house, 1967, 325 pp.
94. Rosenberg, H. M., Phil. Trans. Soc., 1955, Vol. A247, pp. 441–497.
95. Hearmon, R. F. S. Adv. in Phys., 1956, Vol. 5, pp. 323–382.
96. Busch, G. and Güntherhodt, H. J. Sol. Stat. Phys., 1974, Vol. 29, pp. 235–313.
97. Tallon, J. L. and Woffenden, A. J. Phys. Chem. Sol., 1979, Vol. 40, pp. 831–837.
98. Gerlich, D. and Fisher, E. S. J. Phys. Chem. Sol., 1969, Vol. 30, pp. 1197–1205.
99. Vukalovich, M. P., Ivanov, A. I., Fokin, L. R., and Yakovlev, A. T. Thermophysical properties of liquids. Moscow, Nauka, 1970, pp. 135–137.
100. Kokoin, I. K., ed. Tables of physical quantities. Moscow, Atomizdat, 1976, 1006 pp.
101. Reference book on electrical engineering materials. Moscow-Leningrad, Gosenergoizdat, 1960, 511 pp.
102. Gripshover, R. L., van Zytveld, J. B., and Bass, J. Phys. Rev., 1967, Vol. 163, pp. 598–603.
103. Balazyuk, V. N., Geshko, E. I., Mikhail'chenko, V. P., and Shariat, B. M. Fiz. Metal Metalloved, 1976, Vol. 42, pp. 854–859.
104. Filyand, M. A. and Semenova, E. M. Properties of the rare elements. Moscow, Metallurgiya, 1964, 912 pp.
105. Pashaev, B. P., Palachaev, D. K., Paschuk, E. G., and Revelis, V. G. Inzh.-Fiz. Zh., 1980, Vol. 38, pp. 614–620.
106. Olsen-Bär, M. and Powell, R. W. Proc. Roy. Soc., 1951, Vol. A209, pp. 1099–1103.

107. Chandrasekhar, B. S. and Rayne, J. A. Phys. Rev., 1961, pp. 1011–1014.
108. Vold, C. L., Glicksman, M. E., and Kammer, E. W. J. Phys. Chem. Sol., 1977, Vol. 38, pp. 157–160.
109. Kuvandikov, O. K., Cheremushkina, A. V., Vasil'eva, R. P., Fiz. Metal Metalloved, 1972, Vol. 34, pp. 864–869.
110. Goldratt, E. and Greenfield, A. J. J. Phys. F: Met. Phys., 1980, Vol. 10, pp. L95–L99.
111. Ferris, R. W., Shepard, M. L., and Smith, J. F. J. Appl. Phys., 1963, Vol. 34, pp. 768–770.
112. Greenfield, A. J. Phy. Rev., 1964, Vol. 135, pp. 1589–1595.
113. Prokhorenko, V. Ya. Reviews of thermophysical properties of substances: Structure and electrical properties of liquid metals. Moscow, Inst. Vys. Temp. AN SSSR, 1980, No. 6, 120 pp.
114. Kammer, E. W., Cardinal, L. C., Vold, C. V., and Glicksman, M. E. J. Phys. Chem. Sol., 1972, Vol. 33, pp. 1891–1898.
115. Drapkin, B. M., Birfeld, A. A., Kononenko, V. K. and Kalyukin, Yu. N. Fiz. Metal Metalloved, 1980, Vol. 49, pp. 1075–1080.
116. Karamargin, M. C., Reynolds, C. A., and Lipschultz, F. P. Phys. Rev., 1972, Vol. B5, pp. 2856–2863.
117. Case, S. K. and Gueths, J. E. Phys. Rev., 1970, Vol. B2, pp. 3843–3848.
118. Otter, C. and Arles, L. Rev. Int. Hautes Temp. et. Refract., 1978, Vol. 15, pp. 209–219.
119. Zinov'yev, V. E., Baskakova, A. A., Korshunov, I. G. and Zagrebin, L. D. Inzh.-Fiz. Zh., 1973, Vol. 25, pp. 490–494.
120. Waldorf, D. L. and Alers, G. A. J. Appl. Phys., 1962, Vol. 33, pp. 3266–3269.
121. Powell, R. W. and Tye, R. P. Thermodynamic and transport properties of fluids. London, 1958, 182 pp.
122. Cook, J. G., Laubitz, M. J., and van der Meer, M.-P. J. Appl. Phys., 1974, Vol. 45, pp. 510–513.
123. Frantseevich, I. N., Voronov, F. F., and Bakuta, S. A. Elastic constants and moduli of elasticity of metals and nonmetals. Kiev, Naukova dumka, 1982, 286 pp.
124. Maksimyuk, P. A. and Belyaev, A. E. Fiz. Tverd. Tela, 1973, Vol. 15, pp. 2813–1825.
125. Levitskii, Yu. T. and Ivanov, G. A. Fiz. Metal Metalloved, 1969, Vol. 27, pp. 598–602.
126. Maksimyuk, P. A. and Onanenko, A. P. Fiz. Tverd. Tela, 1981, Vol. 23, pp. 589–591.
127. Michenaud, J.-P. and Issi, J. P. J. Phys. C.: Sol. Stat. Phys., 1972, Vol. 5, pp. 3061–3072.
128. Ivanov, G. A. and Levitskii, Yu. T. Fiz. Metal Metalloved, 1967, Vol. 24, pp. 253–259.
129. Hurle, D. T. J. and Weintraub, S. Proc. Phys. Soc., 1960, pp. 163–166.
130. Fal'kovskii, L. A. Usp. Fiz. Nauk, 1968, Vol. 94, pp. 1–41.
131. Issi, J.-P. Austral. J. Phys., 1979, Vol. 32, pp. 585–628.
132. Baskakova, A. A., Zinov'yev, V. E., and Zagrebin, L. D. Inzh.-Fiz. Zh., 1974, Vol. 26, pp. 1058–1061.
133. Shimizu, M., Takahashi, T., and Katsuki, D. J. Phys. Soc. Japan, 1965, Vol. 20, pp. 1192–1203.
134. Ramji, R. R. and Menon, C. S. Sol. Stat. Comm., 1973, Vol. 12, pp. 527–528.
135. Gorecki, T. Mater. Sci. and Ing., 1980, Vol. 43, pp. 225–230.
136. Colvin, R. V. and Arajs, S. J. Appl. Phys., 1963, Vol. 34, pp. 286–290.
137. Spedding, F. H., Cress, D., and Beandry, B. J. L. Less. Comm. Met., Vol. 23, pp. 263–270.
138. Mardon, P. G., Nichols, J. L., Perarce, J. H. and Pool, D. M. Nature, 1961, Vol. 189, pp. 566–570.
139. Zinov'yev, V. E., Chupina, L. I., and Gel'd, P. V. Fiz. Tverd. Tela, 1972, Vol. 14, pp. 2787–2790.
140. Zinov'yev, V. E. and Korshunov, I. G. Reviews of thermophysical properties of substances, Pt. 1: Review of the experimental data. Moscow, Inst. Vys. Temp. AN SSSR, 1978, No. 4, 121 pp.; Pt. 2: Distinctive features of mechanisms of electron and phonon scattering. Moscow, Inst. Vys. Temp. AN SSSR, 1979, No. 4, 119 pp.
141. Mardykin, I. P. in Physicomechanical and thermophysical properties of metals. Moscow, Nauka, 1976, pp. 105–111.

142. Meaden, G. T. and Sze, N. H. J. Less. Comm. Met., 1969, Vol. 19, pp. 444–447.
143. Ross, J. W. and Isaace, L. L. J. Phys. Chem. Sol., 1971, Vol. 32, pp. 747–750.
144. Colvin, R. V. and Arajs, S. J. Appl. Phys., 1963, Vol. 34, pp. 286–290.
145. Savage, S. J., Palmer, S. B., Fort, D., Jordan, R. G., and Jones, D. W. J. Phys. F: Met. Phys., 1980, Vol. 10, pp. 347–352.
146. Smith, J. F. and Gjevre, J. A. J. Appl. Phys., 1960, Vol. 31, pp. 645–647.
147. Novikov, I. I., Kostyukov, V. I., and Filippov, L. P. Izv. AN SSSR, Metally, 1978, No. 4, pp. 89–93.
148. Novikov, I. I. and Mardykin, I. P. Izv. AN SSSR, Metally, 1976, No. 1, pp. 27–30.
149. Novikov, I. I. and Mardykin, I. P. Atomn. Energ., 1976, Vol. 40, No. 1, pp. 63–64.
150. Zinov'yev, V. E. and Kostyukov, V. I., Inzh-Fiz. Zh., 1980, Vol. 34, pp. 1010–1012.
151. Vedernikov, M. V., Burkov, A. T., Dvunitkin, V. G., and Moreva, N. I. J. Less. Comm. Met., 1977, Vol. 52, pp. 221–245.
152. Edvards, L. R., Schafer, J. and Legvold, S. Phys. Rev., 1969, Vol. 188, pp. 1173–1174.
153. Zinov'yev, V. E., Sperelup, V. I., and Gel'd, P. V. Fiz. Tverd. Tela, 1978, Vol. 20, pp. 1677–1681.
154. Pavlov, V. S., and Pankrat'eva, M. I. Izv. VUZ Fiz., 1970, No. 6, pp. 107–109.
155. Banchila, C. N. and Filippov, L. P. Inzh.-Fiz. Zh., 1974, Vol. 27, pp. 68–71.
156. Khomskii, D. I. Usp. Fiz. Nauk, 1979, Vol. 129, pp. 443–483.
157. Rosen, M. Phys. Rev., Vol. 181, pp. 432–435.
158. Voronov, F. F., Goncharova, V. A., and Stal'gorova, O. V. Zh. Eksp. Teor. Fiz., 1979, Vol. 76, pp. 1351–1354.
159. Nicolas, M. and Jerom, D. Sol. Stat. Comm., 1973, Vol. 12, pp. 523–526.
160. Busch, G., Güntherhodt, H.-J., Kunzi, H. U., and Schapback, L. Phys. Lett., 1970, Vol. 31a, pp. 191–192.
161. Mardykin, I. P. and Vertman, A. A. Izv. AN SSSR, Metally, 1972, No. 1, pp. 95–98.
162. Rosen, M. Phys. Rev., 1969, Vol. 180, pp. 540–544.
163. Greiner, J. D., Schiltz, R. J., Tonnies, I. J., et al. J. Appl. Phys., 1973, Vol. 44, pp. 3862–3867.
164. Volkhenstein, N. V., Fedorov, G. V., and Startsev, V. E. Izv. AN SSSR, Fizika, 1964, Vol. 28, pp. 540–545.
165. Gaibulev, F., Regel, A. R., Khusianov, Kh. Fiz. Tverd. Tela, 1969, Vol. 11, pp. 1400–1402.
166. Mardykin, I. P. and Kashin, V. I. Izv, AN SSSR, Metally, 1973, No. 4, pp. 77–80.
167. Ivliev, A. D. and Zinov'yev, A. D. Fiz. Tverd. Tela, 1981, Vol. 23, pp. 1190–1192.
168. Volkhenstein, N. V. and Fedorov, G. V. Fiz. Tverd. Tela, 1965, Vol. 7, pp. 3213–3217.
169. Lenkerri, J. T. and Palmer, S. B. J. Phys. F: Met. Phys., 1977, Vol. 7, pp. 15–22.
170. Volkhenstein, N. V. and Noskova, I. M. Trudy Inst. Fiz. Met. AN SSSR, Sverdlovsk, 1968, No. 27, pp. 130–134.
171. Mardykin, I. P. Teplofiz. Vys. Temp., 1975, Vol. 13, pp. 211–213.
172. Vereshchaka, I. P., Bryukhanov, A. E., and Nemchenko, V. F. Fiz. Metal Metalloved, 1971, Vol. 31, pp. 882–884.
173. Vedernikov, M. V., Kizhaev, S. A., Petrov, A. V. and Moreva, N. I. Fiz. Tverd. Tela, 1975, Vol. 17, pp. 340–342.
174. Kraftmakher, Ya. A. and Pinegina, T. Ya. Phys. Stat. Sol., 1978, Vol. 47a, pp. K81–K82.
175. Arajs, S. and Dumyre, G. R. Z. Naturforsch., 1966, Vol. 21a, pp. 1856–1859.
176. Devyatkova, E. D. and Zhuse, V. P. Fiz. Tverd. Tela, 1964, Vol. 6, pp. 430–435.
177. Rosen, M. Phys. Rev., 1968, Vol. 166, pp. 561–564.
178. Vedernikov, M. V., Burkov, A. G., and Moreva, N. I. Fiz. Tverd. Tela, 1975, Vol. 17, pp. 3100–3101.
179. Cafe, J. T., Zwart, J., and van Zytveld, J. B. J. Phys. F: Met. Phys., 1980, Vol. 10, pp. 669–676.
180. Curry, M. A., Legvold, S., and Spedding, F. H., Phys. Rev., 1960, Vol. 177, pp. 953–954.
181. Keeton, S. C. and Loucks, T. L. Phys. Rev., 1968, pp. 672–678.
182. Zinov'yev, V. E., Mal'gin, A. V., Zinov'eva, G. P., and Epifanova, K. I. Fiz. Metal Metalloved, 1980, Vol. 50, pp. 659–662.

183. Palmer, S. B., Lee, E. W., and Islam, M. N. Proc. Roy. Soc., 1974, Vol. A338, pp. 341–357.
184. Rosen, M. Phys. Rev., 1968, Vol. 174, pp. 504–514.
185. Nellis, W. J. and Legvold, S. J. Appl. Phys., 1969, Vol. 40, pp. 2267–2269.
186. Kasuya, T. In: Magnetism, Vol. 11. Rado, G. T., Suhl, H. eds. New York, Academy Press, 1966.
187. Souse, J. B., Pinto, R. P., Amado, M. M., et al. J. Phys., 1980, Vol. 41, pp. 573–578.
188. Nellis, W. J. and Legvold, S. Phys. Rev., Vol. 180, pp. 581–583.
189. Souse, J. B., Amado, M. M., Pinto, R. P., et al. J. Phys. F: Met. Phys., 1979, Vol. 9, pp. L77–L81.
190. Zinov'yeva, G. P., Gel'd, P. V., Zinov'yev, V. E., and Sperelup, V. I. DAN SSSR, 1980, Vol. 254, pp. 95–97.
191. Amado, M. M., Souse, J. B., Pinheiro, M. F., et al. Portug. Phys., 1980, Vol. 11, pp. 33–39.
192. Zinov'yev, V. E., Sperelup, V. I., Chuprikov, G. E., and Epifanova, K. I., Fiz. Metal Metalloved, 1977, Vol. 44, pp. 182–183.
193. Palmer, S. B. and Lee, E. W. Proc. Roy. Soc., 1972, Vol. 237, pp. 519–543.
194. Boys, D. W. and Legvold, S. Phys. Rev., 1968, Vol. 174, pp. 372–379.
195. Güntherhodt, H.-J., Hausser, E., and Kanzi, H. U. Phys. Lett., 1974, Vol. 48a, pp. 201–202.
196. Nikitin, S. A., Nakavi, S. M., Solomkin, I. K., and Chistyakov, O. D. Fiz. Tverd. Tela, 1977, Vol. 19, pp. 1301–1308.
197. Salama, K., Brotzen, F. R., and Donoho, P. L. J. Appl. Phys., 1973, Vol. 44, pp. 180–184.
198. Rosen, M., Kalir, D., and Klimker, H. J. Phys. Chem. Sol. 1974, Vol. 35, pp. 1333–1339.
199. Dvunitkin, V. G., Vedernikov, M. V., and Moreva, N. I. Fiz. Metal Metalloved, 1978, Vol. 45, pp. 894–896.
200. Rosen, M. J. Phys. Chem. Sol., 1971, Vol. 32, pp. 2351–2356.
201. Edwards, L. R. and Legvold, S. Phys. Rev., 1968, Vol. 176, pp. 753–760.
202. Dvunitkin, V. G., Vedernikov, M. V., and Moreva, N. I. Fiz. Tverd. Tela, 1979, Vol. 21, pp. 2177–2179.
203. Reandry, B. J., Gschneider, K. A. Sol. Stat. Comm., 1974, Vol. 15, pp. 791–793.
204. Tonniec, J. J., Gschneider, K. A., and Spedding, F. H. J. Appl. Phys., 1971, Vol. 42, pp. 3275–3283.
205. Kanevskii, I. N., Nisnevich, M. M., and Spasskaya, A. A. Izv. AN SSSR, Metally, 1979, No. 3, pp. 210–213.
206. Edwards, L. R., Schaefer, J., and Legvold, S. Phys. Rev., 1969, Vol. 188, pp. 1173–1174.
207. Fischer, E. S. and Renken, C. J. Phys. Rev., 1964, Vol. 135, pp. 482–494.
208. Yeremenko, V. I. Titanium and its alloys. Kiev, Naukovo dumka, 1960, 500 pp.
209. Grimvall, G. Rev. Int. Hautes Temp. et Refract., 1979, Vol. 16, pp. 411–412.
210. Seydel, V. and Fucke, W. J. Phys. F: Met. Phys., 1980, Vol. 10, pp. L203–L206.
211. Bereznikova, N. V., Borzyak, A. I., Lepeshkin, Yu. D., et al. In: Physicomechanical and thermophysical properties of metals. Moscow, Nauka, 1976, pp. 13–21.
212. Powell, R. Usp. Fiz. Nauk, 1971, Vol. 105, pp. 329–350.
213. Vavra, G. Izv. AN SSSR, Metally, 1978, No. 6, pp. 219–220.
214. Padell, A. and Groff, A. J. Nucl. Mat., 1976, Vol. 59, pp. 325–326.
215. Peletskii, V. E. and Bel'skaya, E. A. Electrical resistivity of refractory metals. Moscow, Energiya, 1981, 94 pp.
216. Borzyak, A. N., Lepeshkin, Yu. D., Novikov, I. I., and Uspayeva, N. V. in Physicomechanical and thermophysical properties of metals. Moscow, Nauka, 1976, pp. 59–80.
217. Filippov, L. P. and Yurchak, R. P. Inzh.-Fiz. Zh., 1971, No. 11, pp. 561–577.
218. Arutyunov, A. V., Banchila, S. N., and Filippov, L. P. Teplofiz. Vys. Temp., 1972, Vol. 10, pp. 425–428.
219. Montaque, S. A., Draper, C. W., and Rosenblatt, G. M. J. Phys. Chem. Sol., 1979, Vol. 40, pp. 987–992.
220. Walker, E. Sol. Stat. Comm., 1978, Vol. 28, pp. 587–589.

221. Ashkenazi, J., Dacorogna, M., and Peter, M. Phys. Rev., 1978, Vol. B18, pp. 4120–4131.
222. Walker, E. and Bujard, P. Sol. Stat. Comm., 1980, Vol. 34, pp. 691–693.
223. Kashtalyan, Yu. A. Elastic parameters of materials at high temperatures. Kiev, Naukova dumka, 1980, 112 pp.
224. Farraro, R. J. and McLellan, Rex B. Met. Trans., 1979, Vol. 10A, pp. 1699–1702.
225. Kuritnyk, I. P., Stadnyk, B. I., and Kuritnyk, T. I. Electronic engineering materials, 1975, No. 2, pp. 49–53.
226. Peletskii, V. E., Timrot, D. L., and Voskresenskii, V. Yu. High temperature investigations of thermal and electrical conductivity of solids. Moscow, Energiya, 1971, 192 pp.; Peletskii, V. I., Amasovich, E. S. Teplofiz. Vys. Temp., 1977, Vol. 15, pp. 1202–1208.
227. Lebedev, S. V., Savvatimskii, A. I., and Smirnov, Yu. B. Zh. Eksp. Teor. Fiz., 1972, Vol. 42, pp. 1752–1760.
228. Kraev, O. A. and Stel'makh, A. A. in Investigations at high temperatures. Novosibirsk, Nauka, 1966, pp. 55–74.
229. Palmer, S. B. and Lee, E. W. Phil. Mag., 1971, Vol. 24, pp. 311–318.
230. Street, R., Munday, B. C., Window, B., and Williams, I. R. J. Appl. Phys., 1968, Vol. 39, pp. 1050–1059.
231. Goff, J. F. Phys. Rev., 1970, Vol. B1, pp. 1351–1352.
232. Meaden, G. T., Rao, K. V., and Tee, K. T. Phys. Rev. Lett., 1970, Vol. 25, pp. 359–362.
233. Moore, J. P., Williams, R. K., and McElroy, D. L. Phys. Lett. Rev., 1970, Vol. 24, pp. 587–588.
234. Akida, C. and Mitsui, T. J. Phys. Soc. Japan, 1972, Vol. 32, pp. 644–652.
235. Anderson, J. M., Stewart, A. D., and Ramsay, I. Phys. Stat. Sol., 1970, Vol. 37, pp. 325–328.
236. Stewart, A. D. and Anderson, J. M. Phys. Stat. Sol., 1971, Vol. 45b, pp. K89–K93.
237. Baum, V. A., Gel'd, P. V., and Suchilnikov, S. I., Izv. AN SSSR, Metallurgiya i Gornoe Delo, 1964, No. 2, pp. 149–155.
238. Farraro, R. and McClellan, R. B. Metal. Trans., 1977, Vol. 84, pp. 1563–1565.
239. Voronov, F. F., Prokhorov, V. M., Gromytskii, E. L., and Ilina, G. G. Fiz. Metal Metalloved, 1978, Vol. 45, pp. 1263–1267.
240. Peletskii, V. E. Teplofiz. Vys. Temp., 1976, Vol. 14, pp. 522–527.
241. Hoch, M. Thermodyn. Nucl. Mater., 1974. Proc. Symp., Vol. 2. Viena, 1975, pp. 113–121.
242. Peletskii, V. E. High Temper.—High Press., 1970, Vol. 2, pp. 167–170.
243. Hust, J. G. High Temper.—High Press., 1976, Vol. 8, pp. 377–381.
244. Peletskii, V. E. Teplofiz. Vys. Temp., 1977, Vol. 15, pp. 728–734.
245. Rosen, M. Phys. Rev., 1968, Vol. 165, pp. 357–359.
246. Salli, A. Manganese. Moscow, Metallurgiya, 1959, 296 pp.
247. Mozi, N. J. Phys. Soc. Japan, 1974, Vol. 37, pp. 1285–1290.
248. Akshentsev, Yu. N., Baum, B. A., and Gel'd, P. V. Izv. AN SSSR, Metally, 1969, No. 4, pp. 177–181.
249. Love, G. R., Koch, C. C., Whally, H. L., and McNutt, Z. R. J. Less. Comm. Met., 1970, Vol. 20, pp. 73–75.
250. Spitsyn, V. I., Zinov'yev, V. E., Gel'd, P. V., and Balakhovskii, O. A., DAN SSSR, 1975, Vol. 221, No. 1, pp. 145–148.
251. Koch, C. C. and Love, G. R. J. Less. Comm. Met., 1967, Vol. 12, pp. 29–34.
252. Baker, D. E. J. Less. Comm. Met., Vol. 8, pp. 435–439.
253. Volkhenstein, N. V., Startsev, V. E., Cherepanov, V. I., et al. Fiz. Metal Metalloved, 1978, Vol. 45, pp. 1187–1199.
254. Arutyunov, A. V. and Filippov, L. P. Teplofiz. Vys. Temp., 1970, Vol. 8, pp. 1095–1097.
255. Peletskii, V. E. In: Investigations and applications of rhenium alloys. Moscow, Nauka, 1975, 75 pp.
256. Kamm, G. N. and Anderson, J. R. Phys. Rev., 1970, Vol. B2, pp. 2944–2954.
257. Zinov'yev, V. E., Savitskii, E. M., Gel'd, P. V., et al. Teplofiz. Vys. Temp., 1978, Vol. 16, pp. 971–977.

258. Volkhenstein, N. V., Dyakina, V. P., Startsev, V. E., et al. Fiz. Metal Metalloved, 1974, Vol. 38, pp. 718–737.
259. Powell, R. W., Tye, R. P., and Woodman, M. J. J. Less. Comm. Met., 1967, Vol. 12, pp. 1–10.
260. Savitskii, E. M., Gel'd, P. V., Zinov'yev, V. E., et al. DAN ASSR, 1976, Vol. 229, pp. 841–844.
261. Anderson, O. K. Phys. Rev., 1970, Vol. B2, pp. 883–906.
262. Tramm, M. M. and Smith, N. Phys. Rev., 1974, Vol. B9, pp. 1353–1364.
263. Rudnitskii, A. A. In: Thermoelectric properties of noble metals and alloys. Moscow, AN SSSR publishing house, 1956, 143 pp.
264. MacFarlane, R. E., Rayne, J. A., and Jobes, C. K. Phys. Lett., 1966, Vol. 20, pp. 234–235.
265. Gel'd, P. V. and Zinov'yev, V. E. Teplofiz. Vys. Temp., 1972, Vol. 10, pp. 656–657.
266. Vuillemen, J. J. Phys. Rev., 1965, Vol. 144, pp. 397–405.
267. Mueller, F. M., Freeman, A. J., and Dimmock, J. O. Phys. Rev., 1970, Vol. B1, pp. 4617–4635.
268. Hoare, F. E., Mathhews, J. C., and Walling, J. C. Proc. Roy. Soc., 1963, Vol. A216, pp. 502–515.
269. Budworth, D. W., Hoare, F. E., and Preston, J. Proc. Roy. Soc., Vol. A257, pp. 250–262.
270. Narth, A. J. Appl. Phys., 1968, Vol. 39, pp. 553–555.
271. Walker, E., Ortelli, J., and Peter, M. Phys. Lett., 1970, Vol. 31A, pp. 240–241.
272. Wenemann, C. and Steinmann, S. Sol. Stat. Comm., 1974, Vol. 15, pp. 281–285.
273. Masumoto, H., Saito, H., and Kadovak, S. Science Reports RITU, 1968, Vol. 19A, pp. 294–303.
274. Chupina, L. I., Zinov'ev, V. E., Polyakova, V. P., et al. Fiz. Metal Metalloved, 1979, Vol. 48, pp. 476–483.
275. Fradin, E. Y., Koelling, D. D., Freeman, A. J., and Watson-Yang, T. Phys. Rev., Vol. B12, pp. 3570–3574.
276. Shimizu, M. and Katuski, A. J. Phys. Soc. Japan, 1964, Vol. 19, pp. 1135–1141.
277. MacFarlane, R. E., Rayne, J. A., and Jones, C. K. Phys. Lett., 1965, Vol. 18, pp. 91–92.
278. Zinov'yev, V. E., Krentsis, R. P., and Gel'd, P. V. Fiz. Tverd. Tela, 1968, Vol. 10, pp. 2826–2828.
279. Laubitz, M. J. and Matsumura, T. Can. J. Phys., 1972, Vol. 50, pp. 196–205.
280. Ackerman, M. W., Wu, K. L., and Ho, C. Y. In: Thermal conductivity. New York, 1976, Vol. 14, pp. 46–57.
281. Slask, G. A. J. Appl. Phys., 1964, Vol. 35, pp. 339–344.
282. Fradin, F. Y. Phys. Rev. Lett., 1974, Vol. 33, pp. 158–161.
283. Flynn, D. R. and O'Hagan, M. E. J. Res. Nat. Bur. Stand., 1967, Vol. 71c, pp. 255–261.
284. Powell, R. W. and Tye, R. W. Brit. J. Appl. Phy., 1963, Vol. 14, pp. 662–666.
285. Narath, A. and Weaver, H. T. J. Appl. Phys., 1970, Vol. 41, pp. 1077–1078.
286. Kraftmakher, Ya. A. and Sushkova, G. G. Fiz. Tverd. Tela, 1974, Vol. 16, pp. 138–142.
287. Yasui, M., Hayashi, E., and Shimizu, M. J. Phys. Soc. Japan, 1973, Vol. 34, pp. 396–403.
288. Dever, D. J. J. Appl. Phys., 1972, Vol. 43, pp. 3293–3301.
289. Garber, R. I. and Kovalev, A. I. Zavod Lab. 1958, No. 4, pp. 477–479.
290. Lauchbary, M. D. and Sannders, N. H. J. Phys. F: Met. Phys., 1976, Vol. 6, pp. 1967–1977.
291. Arajs, S. and Colvin, R. V. Phys. Stat. Sol., 1964, Vol. 6, pp. 797–802.
292. Fulkerson, W., Moore, J. P., and McElroy, D. L. J. Appl. Phys., 1966, 2639–2644.
293. Oleinikov, P. P. Teplofiz. Vys. Temp., 1981, Vol. 19, pp. 533–542.
294. Kraftmakher, Ya. A. and Pinegina, T. Yu. Fiz. Tverd. Tela, 1974, Vol. 16, pp. 132–142.
295. Zinov'yev, V. E., Abel'skii, Sh. Sh., Sandakova, M. A., et al. Zh. Eksp. Teor. Fiz., 1972, Vol. 63, pp. 2221–2225.
296. Kraftmakher, Ya. A. and Pinegina, T. Yu. Phys. Stat. Sol., 1970, Vol. 42, pp. K151–K153.
297. Baum, B. A. Metallic liquids. Moscow, Nauka, 1979, 120 pp.
298. Sergeev, O. A. and Chadovich, T. Z. Trudy Metrological Inst. SSSR, 1974, No. 155 (215), pp. 78–86.

299. Ivliev, A. D. and Zinov'ev, V. E. Teplofiz. Vys. Temp., 1980, Vol. 18, pp. 532–539.
300. Khandros, V. O. and Bogolyubov, I. A. Fiz. Metal Metalloved, 1976, Vol. 42, pp. 1322–1324.
301. Laubitz, M. J. and Matsumura, T. Can. J. Phys., 1973, Vol. 51, pp. 1247–1256.
302. Kraftmakher, Ya. A. and Pinegina, T. Yu. Fiz. Tverd. Tela, 1971, Vol. 13, pp. 2799–3001.
303. Greig, D. and Morgan, G. J. Phil. Mag., 1973, Vol. 27, pp. 929–940.
304. Alers, G. A., Neighbours, J. R., and Sato, H. J. Phys. Chem. Sol., 1960, Vol. 13, pp. 40–55.
305. Takahashi, S. and Yamamoto, E. J. Japan Inst.: Metals, 1973, Vol. 37, pp. 373–375.
306. Bel'skaya, E. A. and Peletskii, V. E. Teplofiz. Vys. Temp., 1981, Vol. 19, pp. 525–532.
307. Laubitz, M. J., Matsumura, T., and Kelly, P. I., Canad. J. Phys., 1976, Vol. 56, pp. 92–99.
308. Powell, R. W., Tye, R. P., and Hickman, M. J. Int. J. Heat and Mass Transf., 1965, Vol. 8, pp. 679–685.
309. Papp, E., Szabo, Gy., and Tichy, G. Sol. Stat. Comm., 1977, Vol. 21, pp. 487–490.
310. Long, K. A. Phys. Stat. Sol., 1977, Vol. 79b, pp. 155–165.
311. Greiner, J. D., Peterson, D. T., and Smith, J. F. J. Appl. Phys., 1977, Vol. 48, pp. 3357–3361.
312. Schetter, H. G., Martin, J. J., Schmidt, F. A., and Danielson. G. C. Phys. Rev., 1969, Vol. 187, pp. 801–804.
313. Fisher, E. S. J. Nucl. Mat., 1966, Vol. 18, pp. 39–54.
314. Rosen, M. Phys. Rev., 1968, Vol. 28A, pp. 438–440.
315. Merz, M. D. and Kiarmo, H. E. In: Plutonium—1975 and other actinides. Proc. 5th Int. Conf., Baden-Baden, 1975. Amsterdam c.a. 1976, pp. 106–107.
316. Wagner, P. J. Less. Comm. Met., 1971, Vol. 24, pp. 106–107.
317. Kay, A. E. and Linford, P. F. T. The elastic constants of plutonium. Plutonium 1960, Proc. 2nd Int. Conf., Grenoble, April 1960. 1961, pp. 51–58.
318. Ledbetter, H. M. and Moment, R. L. Acta Metall., 1976, Vol. 24, pp. 891–899.
319. Andrew, J. F. J. Phys. Chem. Sol, 1967, Vol. 28, pp. 577–580.
320. Brodsky, M. B., Arko, A. J., Harvey, A. R., and Nellis, W. J. In: The actinides. Electronic structure and related properties. Freeman, A. J., Darby, J. B., eds. New York, Acad. Press, 1974, Vol. 2, pp. 185–264.
321. Karpenko, B. V. In: Electronic structure and physical properties of rare earths and actinides. Sverdlovsk, Ural. Nauch. Tsentr AN SSSR, 1981, pp. 86–113.
322. Hall, R. O. A. and Lee, J. A. J. Low Temp. Phys., 1971, Vol. 4, pp. 415–420.
323. Andrew J. E. and Klemens, P. G. Therm conduct. Vol. 17. Proc. 17th Int. Conf., Gaithersburg, Md., 15–18 June, 1981. New York, 1983, pp. 209–218.
324. Ivanov, V. V., Lebedev, S. V., and Savvatimskii, A. I. Teplofiz. Vys. Temp., 1983, Vol. 20, pp. 1093–1097.
325. Ivanov, V. V., Lebedev, S. V., and Savvatimskii, A. I. Teplofiz. Vys. Temp., 1983, Vol. 21, pp. 390–392.
326. Zinov'yev, V. E., Ivliev, A. D., Korshunov, I. G., and Il'inykh, S. A. Review of the thermophysical properties of substances: Thermal conductivity and diffusivity of transition metals at high temperatures in the vicinity of phase transformation points. Moscow: Inst. Vys. Temp. AN SSSR, 1982, No. 5, 63 pp.
327. Gel'd, P. V., Il'inykh, S. A., Taluts, S. G., and Zinov'yev, V. E. DAN ASSR, 1982, Vol. 267, pp. 602–604.
328. Il'inykh, S. A., Zinov'yev, V. E., Taluts, S. G., and Gel'd, P. V. Teplofiz. Vys. Temp., 1984, Vol. 22, No. 3.

INDEX